# Praise for

# You Can Count On It

### by 2012 workshop participating teachers

1. As teachers we need to increase our knowledge/understanding of mathematical skills beyond the grades we teach. This allows us to grow the minds of our students beyond grade level expectations.

2. Great week of stretching myself as a thinker! Gave me better understanding of math vs just "knowing how" to do it but not why it works.

3. I became more fluent in math. Also able to bring to the classroom another tool to aid the students when problem solving. … I learned a lot!

4.
   - I felt so respected as I grappled with these math concepts. Thanks, Zab!
   - I particularly liked the Benjamin DVD.
   - I see these rods as very valuable for students who struggle and (those who) are very advanced. Can't wait to use them.

5. Had a great time. Felt very comfortable even though I was out of my element. Almost makes me want to go back to H.S. and redo math course work because I feel I would get it! I will use blocks to teach (+, -) and fractions.

All of the teachers indicated that this institute was "worthy of follow-up."

# You Can Count On It:
## A Mentor's Arithmetic Patterns For Elementary Students

# You Can Count On It:
## A Mentor's Arithmetic Patterns
## For Elementary Students

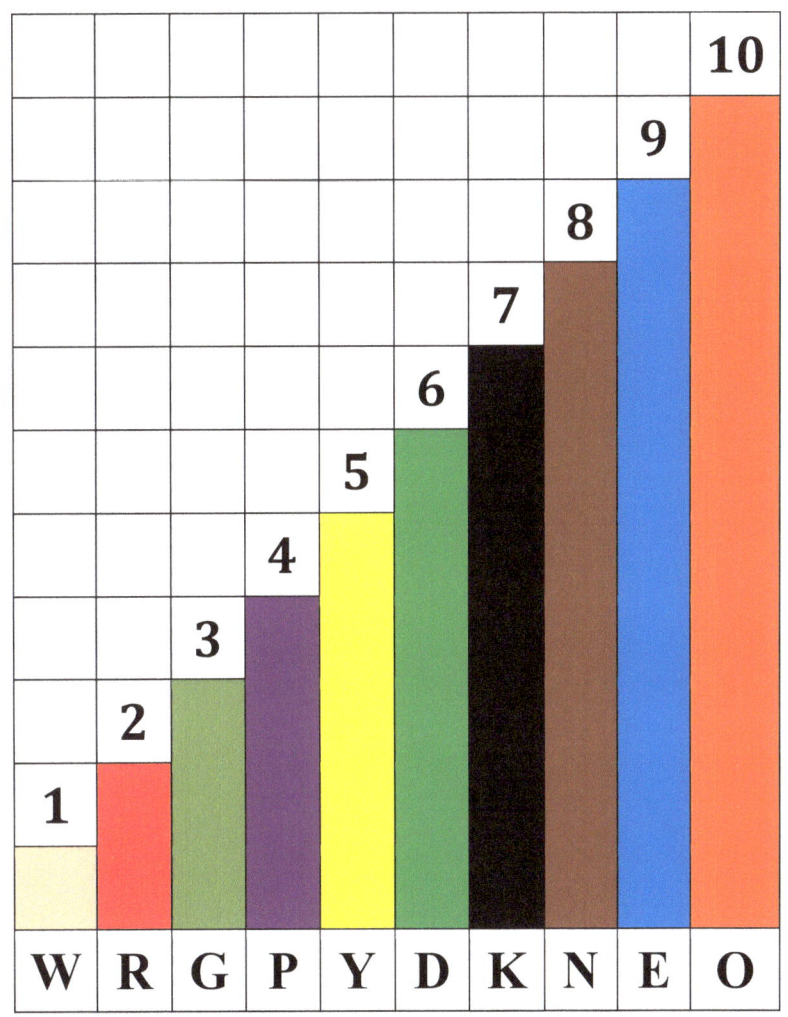

### Alexander Z. Warren

You Can Count On It:
A Mentor's Arithmetic Patterns For Elementary Students

Copyright © 2011 by Alexander Z. Warren
All rights reserved

The name Cuisenaire® and the color sequence of the Rods
(found on the cover of this book)
are registered trademarks of ETA/hand2mind®

ISBN-13: 978-1491037850
ISBN-10: 1491037857

Library of Congress Control Number: 2013915989

CreateSpace Independent Self-publishing Platform
North Charlston, SC

Printed in the United States of America

## Dedication

This book I dedicate to my first five extraordinary elementary students
(E. Benoit, C. Corey, D. Hunt, P. Kirk, M. Macey),
particularly those who participated in my Calculus Summer Camp,
and also to the many Amesbury, Newburyport and Salisbury children
who have taught me how much mathematics they could learn and enjoy.

## Acknowledgements

I want to express my thanks to my former colleague David A. Penner,
whose editorial skills and mathematical insights
have improved the text in so many ways.
Laurence Krenis, MD, also deserves considerable thanks
for his careful reading and challenging remarks.
Thanks also to John Flynn for his kind comments
and careful corrections.
Lucie C. Garnett has the capacity to read the text
as one of her students might. Her suggestions
and her corrections were much appreciated.
Bob Hamilton has made significant and helpful comments
about my text, and I appreciate his insights.
(Any and all failings remain my sole responsibility.)

# Introduction

**Main Purpose:**

Why inflict another arithmetic book on our teachers, when so many already exist? With so many elementary school teachers feeling stressed out with all they are now supposed to do, why ask them to change, to relearn, to re-image? In response, I am drawn to University Professor Helen Vendler's quote in the 2011 Harvard Magazine expressing her concern for the humanities: "What is in place has failed notoriously to make our students eager to read and to learn." I would only change "eager to read" to "eager to do mathematics." I can hear my reader's first response: So, wise guy, what would you have us do? This small book attempts to address that question. It seeks to offer another way, a way that provides the most able a narrowly defined introduction to many traditional topics ending with polynomial calculus, perhaps by the sixth grade, and provides the more slowly moving among us with ideas built on innately human attributes.

Why calculus? Calculus gives its possessors the power to analyze many problems traditional algebra, geometry and trigonometry will not solve. What innately human attributes? The Russian psychologist, L. S. Vygotsky (1896-1934), points to several human strengths over the other primates. People can talk to each other and improve their problem-solving skills by wrestling with each other's ideas. People can work to learn something that does not solve an immediate problem, a problem projecting into the near future. Great apes can use tools, like a long stick to get a banana they want now on the other side of a fence, but cannot learn to solve another problem beyond their immediate view. By posing a variety of problems children usually find hard and showing them they can turn the seemingly impossible into the obvious, we will encourage them to investigate new problems and methods, so they can feel the value of recognizing, understanding and solving many types of problems.

I think of young children at the bottom of the sequence of school years, traveling in a submerged submarine. To take the whole ship to the surface, college and graduate school, will require a long time, dealing, at each level, with the wide breadth of each subject. However, if we make it possible to peek through the periscope, they can acquire a useful view of what lies above, and a better sense of the importance of the intermediate levels.

E.D. Hirsch, Jr. refers to "the so-called Matthew Effect" in his NY Times article (9/18/11), citing the "remarkable linguistic plasticity of young (2-10 years old) minds." Those children fortunate enough to acquire a large vocabulary at this critical age find later learning a natural addition, while those given impoverished verbal experience may lose even the little they understood. I believe this harsh story applies to mathematics as well as to English. So this book pushes for depth ($6^{th}$ grade calculus) over breadth (complete coverage), depth quick learners can view and enjoy in just one hour a week. I hope the students, instructed in this manner, will have the wonderful experience of seeing "impossible" problems turned into "obvious" solutions.

**Primary Assumptions:**

In addition to Vygotsky's observations about the innate human capacities for speech and working toward delayed rewards, I assume several other attributes I believe I have observed: the ability to count, to recognize patterns of many sorts, to distinguish between colors and lengths, to question, to look for and give evidence, and most importantly to play. Illiterate people can count, recognize language patterns, and distinguish among colors. Most importantly, we have to strengthen children's natural inclination to play, to play with ideas, linguistic and mathematical, recognizing how these ideas relate to them and to the world around them.

Why have we asked students to memorize multiplication tables, when they can construct them by counting? Why have we not found appropriate problems to show the power of mathematics, such as how can we measure distances to objects we cannot reach? Why have we left out compass and straight edge geometry constructions that so many future builders and engineers could use daily? I assume children would like to know how things work, how they are put together, and why one approach 'gets there' faster and easier than another. Discussing these various topics leads to enriched learning, superior to the handing out of repetitive problem sheets to keep the students busy. I assume engaging the young mind fosters an engaged older mind, eager to pursue learning.

**Personal History:**

This course has evolved out of a long sequence of events in my life. Most recently, I have been teaching mathematics to a variety of elementary and middle school children from Salisbury, Massachusetts and its neighboring environs. When I met three Salisbury third and fourth graders of great ability eight years ago, I did not know what they could do. They had to teach me! I just knew they lacked any challenge, sleeping in math class, reading novels, and looking at "fillers" and all the while getting top grades. I felt their expectations pushing me to find their limits.

From my sabbatical leave in Salzburg, Austria (1973-74), I learned about Cuisenaire rods, having seen first graders cheer when their teacher announced they were going to have a demonstration math class for a visiting University of Salzburg class. The class demonstrated several exercises found in this book: determining the color of two rods by feeling their lengths and evaluating expressions such as
$$\frac{2}{3} \times 9 + \frac{3}{4} \times 12.$$
Some of the children gave out the answers to the fraction problem without reference to the rods and without delay.

An even earlier experience (1954) in a Swiss teacher training school led me to consider counting as an important and varied activity. Going back even further (1947-48), when I was only 12, I was duly impressed by being handed a yard stick and told to measure the height of the chapel tower. Of all these experiences, along with those I may have forgotten, the on-going search with my Salisbury students for easier and better entrances into mathematics has continually shaped and reshaped my approach to the subject. I hope my reader will not simply accept this play as the final word but accept the challenge to go one better. I would also appreciate hearing from my readers of the ways they think I could improve the presentation.

**Topics and Their Presentation:**

This play gives one possible sequence of topics, but not the only one. When students don't understand a topic repetition may help, or a different entry might be a welcome change. These topics in arithmetic attempt to prepare the student for the same or similar patterns in algebra or elsewhere. The use of some calculators may require the use of parentheses in expressions where the usual order of operations gets changed. The Sieve of Eratosthenes occurs at the end of the book, more because of the many pages of tables than because of the required preparation. Children starting this course from different grades with diverse backgrounds may require changing gears to keep them feeling they are constantly learning something new and having fun exploring the topics.

**Atmospherics**

I have only in the past six years felt completely free to create a distinctive atmosphere for my lessons, a mood created by softly playing baroque music during the class and a playful paper-folding activity at the end of a tough lesson. Colin Rose wrote a book, <u>Accelerated Learning</u>, pointing out the use of left-hand and right-hand brain activity during the usual classes, usually heavily balanced toward the left. The patterned music (Bach, Corelli, Handel, Telemann, Vivaldi) includes, in a productive way, the other side, releasing tension and producing a relaxed learning atmosphere. I carry a bag of CD's, a Walkman and a small up-to-date sound system and always use it. At the end of a period, I try to lighten up a serious conversation by making my own variety of paper airplanes. I have discovered the children get excited and run around after so much sitting. They also display a distressing awkwardness in "simple" manipulative activities. This would not qualify as manual arts, but shows our children's lack of dexterity with activities other than those involving a keyboard.

**Conversational Method:**

I describe my method as conversational. As a mentor, I have to assume different roles: behavior police, enthusiasm cheerleader, problem poser and argument instigator among others. I want to have the students carry as much of the burden of explaining, correcting and asking as possible. Can they see the relationship to themselves of the topic under discussion? Can they offer evidence their point of view, or answer, is the best or at least correct? The use of Cuisenaire rods has a central role to play in making abstract objects like numbers into concrete ones like colored wooden blocks. To promote students' capacity to argue their cases, I try to go so far as to tell them I think 4 divides into 59 evenly, something that we all know is incorrect. But I want to hear students force me to back off that statement and to accept 59's quality as a prime. They gain confidence and persuasiveness by having to back me into a corner, learning as they do some of the standard forms of argumentation. I want them to know how to construct the multiplication tables, as addition tables with most of the sums left out. They should early on learn the names of various types of numbers, learning how to compute with all sorts of combinations of operations. I do not here offer the method of calculating long division results or square root approximations, but try to make the idea of division and root taking clear by giving students examples they can handle. With very able students, I try to change the approach to problems by turning the problem around or upside down as soon as they appear to have the idea. Alternative approaches should give students choices they might make to find easier solution processes. Again, please look for methods I have failed to display or even to know about. So you and your students will enjoy the experience, play math!

AZW
Newburyport, MA
October 30, 2011

# You Can Count On It

## Table of Contents

**Act 1**  Forming the Basis
- Scene 1  Introducing Cuisenaire Rods  1
- Scene 2  Finding the Missing Rod  3
- Scene 3  Two-Rod Trains beyond Orange  4
- Scene 4  The Color of the Difference  5
- Scene 5  Guessing the Colors from Lengths  8
- Scene 6  Counting in Colors  9

**Act 2**  Prime or Composite
- Scene 1  Sums to Five  10
- Scene 2  Sums to Six  12
- Scene 3  Black  14
- Scene 4  Brown  16
- Scene 5  Blue and Orange  17
- Scene 6  Red, Green, or Purple  18

**Act 3**  Counting
- Scene 1  Learn a Little; Gain a Lot  19
- Scene 2  Counting by 2's to Twenty  21
- Scene 3  Cuisenaire Color Coding  23
- Scene 4  Counting by Fours to Forty  25
- Scene 5  Counting by Sixes to Sixty  29
- Scene 6  Counting by 8's  32
- Scene 7  Counting by 3's, An Odd Job  34
- Scene 8  Counting by 5's  36
- Scene 9  Surviving the 7's  37
- Scene 10  Counting by 9's  39
- Scene 11  Collecting Counting  40
- Scene 12  Identifying Factors  44

**Act 4**       Operations on Numbers
      Scene 1      Squares and Their Roots      51
      Scene 2      Rectangles      54
      Scene 3      Divide and Conquer      55
      Scene 4      Lattice Multiplication      56
      Scene 5      Conditional Replacement      60
      Scene 6      Reaching the remainders      62
      Scene 7      Clocks      65
      Scene 8      Modular Arithmetic      68


**Act 5**       Fractions and Exponents
      Scene 1      Fractional Parts      70
      Scene 2      Mixed Operations      72
      Scene 3      Mixed Operations with Fractions      73
      Scene 4      Yet Another Multiplication!      74
      Scene 5      Consecutive Numbers      76
      Scene 6      Exponents      79
      Scene 7      Mixed Operations with Exponents      80
      Scene 8      Changing Precedence      81


**Act 6**       Primes, LCM's, and GCD's
      Scene 1      Prime Numbers Revisited      83
      Scene 2      One-color to One-class Trains      85
      Scene 3      Prime Numbers' Reduced Test      87
      Scene 4      LCM      91
      Scene 5      Euclidean Algorithm & GCD      92


**Act 7**       Areas and Parentheses
      Scene 1      Rectangles      93
      Scene 2      Difference of Two Squares      95
      Scene 3      Ross' Rule      98
      Scene 4      Adding Areas for New formulas      99
      Scene 5      The Difference of Two Areas      102
      Scene 6      Multiplying Two Sums      104
      Scene 7      Order of Subtractions      106
      Scene 8      Multiplying Differences      107
      Scene 9      A Difference Times a Sum      109

| **Act 8** | | Directed Numbers | |
|---|---|---|---|
| | Scene 1 | Number Line | 113 |
| | Scene 2 | Combining Negative Numbers | 116 |
| | Scene 3 | Areas with Negative Numbers | 118 |
| | Scene 4 | More Negative Expressions | 119 |

| **Act 9** | | Rational Numbers | |
|---|---|---|---|
| | Scene 1 | Common Denominators | 120 |
| | Scene 2 | The Least Common Denominator | 123 |
| | Scene 3 | Multiple Additions & Subtractions | 125 |
| | Scene 4 | Multiplying and Dividing | 130 |
| | Scene 5 | Four Operations Together | 133 |

| **Act 10** | | Exponents, Components | |
|---|---|---|---|
| | Scene 1 | Exponents | 136 |
| | Scene 2 | Exponents as Components | 138 |
| | Scene 3 | Difference of Two Cubes & Beyond | 139 |
| | Scene 4 | Sieve of Eratosthenes | 142 |
| | Scene 5 | Divisibility Test for 7 | 147 |
| | Scene 6 | Summing Some Sequences | 149 |

| | |
|---|---|
| Before and After Text | 151 |
| Bibliography | 152 |

# You Can Count On It:
## A Mentor's Arithmetic Patterns
## For Elementary Students

### Alexander Z. Warren

## Act 1         Forming The Basis
(Building a wealth of diverse arithmetic experiences)


### Scene 1         Introducing Cuisenaire Rods
(A classroom with Mentor and young students)

Mentor: We are going to learn math by playing with small colored blocks called Cuisenaire Rods. You should have fun playing with the rods while learning their colors and lengths. We do not consider throwing, eating or losing them playing, but encourage you to make steps or other block constructions. After the math class, make sure the same rods you got out at the start get back into their container.

(The Mentor gives the students time to get the rods out and start playing with them. Many will make stairs out of their collections without being encouraged to do so.)

Mentor: I want everybody to make a set of stair steps, starting with the smallest blocks and ending with the longest. Each step should extend one white rod above the last, until we reach the longest, the orange rods. We want to learn the colors of each size block and how to spell the color names.
        What is the color of the smallest rods?
Students: White.

Mentor: Can anybody spell the word "white?" (Mary raises her hand.)
Mary: W H I T E

Mentor: Let's all write the number "1" and the word "WHITE" on our papers.
(Writing: "1     WHITE" on the board)
        What is the first letter in the word "WHITE?"
Students: W
Mentor: Good. Now we will add the letter "W" to the end of this line.
(Making: "1     WHITE        W" on the board)
Tell me the color of the next smallest rod.
Students: Red.
Mentor: How do you spell red? Frank.
Frank: R E D

Mentor: Good, Frank. How many white rods would make a "train" as long as a red rod?
Frank: 2
Mentor: Now write the number 2, the word "RED" and "R" on a second line.
Writing:       1       WHITE       W
                    2       RED          R

(The Mentor takes the class through all ten colors, noting the problem of using the common first letter, B, in Black, Brown, and Blue to distinguish between them, and therefore switching to the last letters: K, N, and E respectively.)

| | | |
|---|---|---|
| 1 | WHITE | W |
| 2 | RED | R |
| 3 | GREEN | G |
| 4 | PURPLE | P |
| 5 | YELLOW | Y |
| 6 | DARK GREEN | D |
| 7 | BLACK | K |
| 8 | BROWN | N |
| 9 | BLUE | E |
| 10 | ORANGE | O |

Mentor: Class, did you learn something new?
Class: Yes.
Mentor: What did you learn?
Sarah: How to spell yellow.
Tom: The different colors of the rods.
Mentor: Class, did you have some fun?
Class: Yes.
(If they don't respond positively, the Mentor should try to find out why!)


**Between the Scenes**
(Each of these should be repeated, building experience, and preparing for Scene 2.)
1.   Have two students each pick a rod and a third find a rod that will make the shorter 'train' equal to the longer.
2.   Have one student pick a number and another say its color looking away from the rods.
3.   Have one student name a color and another identify the number with this color.
4.   Two students each pick a rod for a train, and a third finds one or two different rods making a train of the same length.

**Act 1     Scene 2     Finding The Missing Rod**

Mentor: We have a variety of games or challenges that help us learn our arithmetic by learning the colors of the rods. When we place one or more rods together end-to-end, we call the sequence a **train**. When two trains have the same length, we say they are equal. Placing one train parallel to another makes it easy to see and feel the equality.

| R | R | G | G | G |
|---|---|---|---|---|
| Y | Y | Y | Y | Y |

We describe this pair of trains of equal length by writing the following **equation**.

$$R + G = Y$$

We read this equation: R plus G equals Y. I will say two colors and you will tell me the color of a single rod equal to that train, or **sum**. Tom, red plus purple?
Tom: Dark green.
Mentor: What equation describes this result?
Tom: R + P = D.

Mentor: White plus black, Sarah?
Sarah: Brown, and the equation: W + K = N.
Mentor: George, brown plus white?
George: N + W = E.
(The Mentor gives all the children a pair of colors to "add," asking them to show their parallel trains and writing the appropriate equation.)

Mentor: Class, did you learn something new?     Class: Yes. How to write equations.
Mentor: Class, did you have some fun?     Class: Yes.
(If they don't respond positively, the Mentor should try to find out why!)


**Between the Scenes**
Two students each pick a rod shorter than the blue one to make a train and a third student finds a rod (or two if necessary) to make a train of equal length. If the original train is longer than an orange rod, the third student should start with an orange rod. Everybody writes the equation describing the two trains.

**Act 1    Scene 3    Two-rod Trains Beyond Orange**

Mentor: Sometimes when we make a train of two rods, we wind up with a train longer than an orange rod. In this case we make a train of one orange rod and an additional color to equal the sum. Mary, what color rod would you have to add to an orange rod to make a train equal to a yellow plus black train?

Mary: I have to add a red rod.
Mentor: Exactly! And what equation expresses this equality?
Mary: Y + K = O + R
Mentor: You've got that, too! What color must we add to the orange to make a train equal to a green plus brown train? Tom.

Tom: That would be red, also.
Mentor: Show us your trains.

Tom: I missed it. I need a white rod.
Mentor: Fine. We always want to be able to give evidence of our calculations. Build the trains! Now try matching blue plus blue. Alexander.
Alexander: I have to add a brown rod to match the train with two blue rods. See my evidence!

Mentor: Alexander, you have demonstrated your conclusion. How would you write the equation expressing this equality?
Alexander: E + E = O + N
Mentor: Excellent.
(The Mentor should make sure everybody can complete a version of this exercise.)

Mentor: Class, did you learn something new?     Yes. Why or why not?
Mentor: Class, did you have some fun?     Yes. If not, why not?

**Act 1        Scene 4        The Color Of The Difference**

Mentor: Now I want you to tell me what color rod will make the shorter train below the same length as the longer.

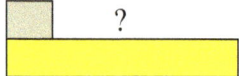

George: Purple!
Mentor: Correct! Now show us your two parallel trains.

Mentor: Now we want to record our findings. Can you tell me how to write an equation expressing this relation by filling in the box with the appropriate letter?

$$W \;+\; \boxed{\phantom{P}} \;=\; Y$$

(All the students write the equation with the letter P in the box.)

$$W \;+\; \boxed{P} \;=\; Y$$

Mary: I wrote: Y = P + W     Is that OK?
Mentor: Yes, that is fine. We note that the white rod is on the left side of the purple one, so we prefer George's Y = W + P. Perhaps for your equation we should turn the two shorter rods around, so the purple rod is on the left and the white one on the right.
       Another way to look at this situation suggests **subtraction**. Suppose we cut the yellow rod into two pieces, the one a centimeter long, the other left-over equals a purple rod. We have subtracted white from yellow, and we can express this operation with the following equation.

$$Y \;-\; W \;=\; \boxed{P}$$

Here we have a new problem. Find the color of the missing rod in the following pair of trains. In other words, subtract green from orange.

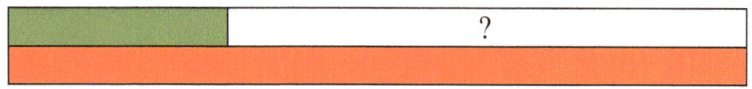

Tom: Dark green!
Mentor: Show me!

(Tom fits the dark green rod to the end of the light green to get the following configuration.)

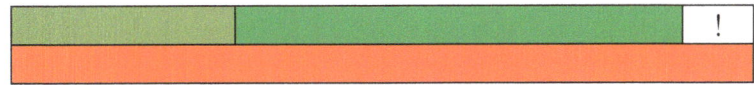

Mentor: Tom, have you found the right rod?
Tom: No. It has to be longer, but not much. How about black? Yes, that works!

Mentor: Good work. What equation expresses your result?
Tom: G + K = O.
Mentor: Fine. You want to visualize these trains, so you can "see" the appropriate rod immediately. Here at the beginning, I want each of you to show the class and me that your guess suits the situation. Later you will be able to picture the right combinations in your head, just imagining the rods.
    Sarah, find the color of the missing rod in the following pair of trains.

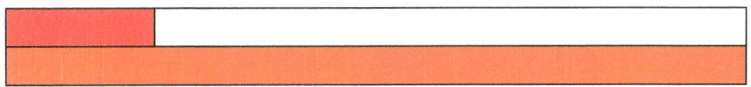

Sarah: Brown!
Mentor: Show me.
(Sarah fills in the blank space with a brown rod.)

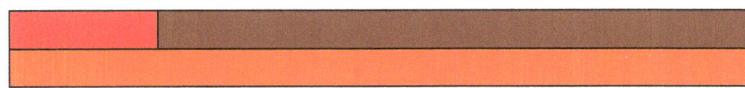

Now tell us how to write this result.
Sarah writes and says: O = R + N or O – R = N.
Mentor: Good. You notice that you can always use the trains to show the correctness of your ideas. Now find the color of this missing rod to make these two trains equal. (Subtract purple from orange) Harry.

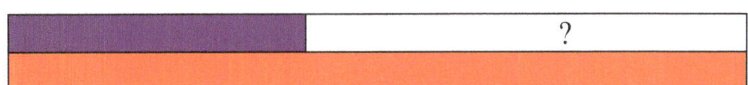

Harry: Black!
Mentor: Show me!

(Harry tries a black rod to discover the purple-black train reaches beyond the orange train. He goes back to the rod collection and pulls out a dark green one. It fits!)

Mentor: Harry, you made a good correction to your first guess. Now dictate the equation for us, so we can record this pair of trains.

Harry writes and says: P + D = O, or
O = P + D, or
O − P = D.

Mentor: Excellent!
(Scene 4 ends after the mentor has given each student the opportunity to create trains of equal length, justifying their decision by placing the appropriate block on the end of the shorter train and dictating his or her equation. The Mentor should repeat this scene until all the students find the appropriate rods confidently.)

Mentor: Class, did you learn something new?   Yes. (Or why not?)
Mentor: Class, did you have some fun?   Yes. (Why or why not?)

**Between the Scenes**
In an attempt to get some random choices, have two students pick rods, without knowing what the other chose, and everybody writes the equation expressing the difference in the two rods, such as Y − W = P. (If both students pick the same color rod, we say the difference equals 0 for which we do not have a rod or a color. To avoid writing the number 0 we could just write one rod equals the other rod. For instance: Y = Y instead of Y − Y = 0.)

**Act 1     Scene 5     Guessing Colors from Lengths**

Mentor: Our game today involves guessing the colors of rods without seeing them, but rather by feeling their lengths in your hands behind your backs. Sarah, please come to the front of the class, face the class with your hands behind you. I will place two unequal rods in your hands and you will tell me their colors.
(The Mentor places a red and a green rod in Sarah's hands and gives her time to figure the colors.)
Sarah: Red and green!

Mentor: Good! Alexander, come try.
(Placing a red and a purple rod in his hands, then waiting.)
Alexander: Red and green!
(Showing the class, experiencing embarrassment, and throwing the blocks on the Mentor's desk, Alexander starts back to his seat.)

Mentor: Alexander, please put those two rods next to each other and tell us what color rod would make the shorter train the same length as the longer rod. Also, ask yourself why you guessed the wrong color of the purple rod. We often learn more from our mistakes than from our right answers, so I want you to think about how you could avoid making this mistake again. Note that the purple rod equals a train of two red rods. Next time you might measure the red block against the purple rod and feel that a train of two of them equals the purple rod. (The Mentor shows the suggested action.) You might also find that you can use your little finger to measure the purple rod. Try measuring the purple block each of these ways.
   Try to learn from your mistakes by looking at the rods, by seeing the relationships, by imagining them in your hands where you cannot see them. Again, don't run away from a mistake without having learned something from it. We all make mistakes, but the best students learn from their errors.

(The Mentor should give each child a chance to guess the colors of two unseen rods, starting with the shorter ones first until the students catch on.)

| | |
|---|---|
| Mentor: Class, did you learn something new? | Yes.   What? |
| Mentor: Class, did you have some fun? | Yes.   (Or why not?) |
| Mentor: Was it sometimes frustrating? | Yes. |
| Mentor: But, in the end? | It was fun? |

**Act 1        Scene 6        Counting In Colors**

Mentor: We have played several games with the rods, hoping that you will remember them both by their lengths and by their colors. Perhaps, for many of you, you now know the colors well enough to tell us the colors of the rods from the shortest to the longest without looking at them. Who thinks he or she can do this?
(Sarah raises her hand.)
Sarah, come stand in front of the class and face the whiteboard. Then count for us in colors!
(Sarah positions herself and recites.)
Sarah: White, red, green, purple, yellow, dark green, black, brown, blue, orange.
Mentor: Good work!
(Sarah takes her seat, and the Mentor gives as many as want to the opportunity to try their memory. The Mentor might try a variation by having the students count backwards or just use the letters.)

Mentor: Can anybody tell us every other color, starting with white? Jane.
Jane: White, green, yellow, black, blue.
Mentor: Someone want to tell us every other color, starting with red? Tom.
Tom: Red, purple, dark green, brown, orange.
Mentor: Excellent. So far, we have concentrated on learning the colors and making equal trains. However, you will recall from our first scene that the colors represent numbers. Red is the color of the number 2. Green is the color of the number 3. In the next scene, we will ask you to practice connecting the colors and the numbers in specific ways.

**Act 2       Prime Or Composite**

**Scene 1       Sums To Five**

Mentor: Now I want you to find the different ways we can make trains the same length as the yellow rod. You should find more than one way to do this. For instance, if I make a train with a red and a green rod, it will equal the yellow rod. Can you find other possibilities? Keep your trains side-by-side, so we can all see what everybody discovers.

(Sarah raises her hand, indicating her eagerness to show her solutions to the problem. She has the trains laid out as shown below.)

Mentor: Sarah, tell us what you have found.
Sarah: A red and green train equals a yellow rod. So does a purple and a white train. I also found that two reds and a white make a train equal to a yellow rod.

Mentor: You have done well. Can you tell us how to record your findings?
Sarah: Yes!
(She reads off her list.)

Y  =  R  +  G
Y  =  W  +  P
Y  =  R  +  W  +  R

Mentor: You did that well, Sarah. Now can anybody else tell me some more trains we might add to Sarah's list? George.

George: I found two trains not on Sarah's list.

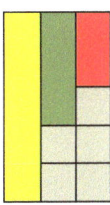

Mentor: How would you record your findings? Write them down and say them to us as you write.
(George says as he writes.)

Y = G + W + W
Y = R + W + W + W

Mentor: Has anybody found yet some other trains the same length as the yellow rod?
Sarah: How about another yellow?
Mentor: Yes. All the yellow rods equal each other. How about another different train? Alexander!
Alexander: I have five white rods in one train.
Mentor: Excellent! Let's look at all the different yellow-length trains we have found.

Mentor: How many trains have we found with only one color (**a one-color train**)?
Alexander: Two. The white train and the yellow train.
Mentor: Can anybody find another one-color train the same length as yellow?
(Silence) Notice that we can try to make a red train the same length as the yellow rod, but a train of two red blocks falls short and a train of three extends too far.

Mentor: Similarly we notice that the green and purple rods create trains too short or too long. So we have used up all the colors of rods shorter than the yellow ones. Because we have just two one-color trains (trains of one color only!) the same length as yellow—one made up solely of whites and the other a yellow—we say that, yellow (or 5) is a **prime** number. In the next scene we look at the dark green rod, asking how many different trains we can make of length equal to dark green and how many one-color trains equal to dark green?

**Act 2     Scene 2     Sums To Six**

Mentor: In this segment, we will look at the dark green rods, following the pattern we saw in Sums to Five. So I want you to find out how many different ways we can make trains the same length as a dark green rod.
(Waiting for the students to make their parallel trains.)
   George, tell us what trains you have made that equal a dark green rod.
George: Purple plus red, two green rods, six white rods, a yellow and a white.

Mentor: Excellent, George. Has George given us all the possibilities? Tom.
Tom: I found three more trains: one with three red rods, another with a red and a green and a white, and a third with a purple and two white rods.
Mentor: Excellent. Let's add those to our display.

Mentor: Do we have any more possibilities? Mary.
Mary: I have two reds and two whites, a green and three whites.
Mentor: Fine. Let's add these to our display.

Mentor: Now we see how many different trains we can think of at our first try. You will be able to find some more if you think about it, but I want you to contemplate the one-color trains we have just made, set off to the right by a blank column. How many have we shown, and what colors do we see? Alexander.

Alexander: I see three one-color trains: green, red, and white.
Mentor: Can anybody find one that is missing?
Alexander: Dark green.
Mentor: Excellent. So now we can count four different one-color trains equal to a dark green rod. We say dark green (or 6) is a **composite** number. Dark green has more than two one-color trains equal to it. We call any number **even**, if we can construct a one-color red train of the same length. We call the rest of the numbers **odd**. Note that yellow (or 5) is odd because we have to add one white rod to two red ones to make a train of the same length. If all red trains are too short or too long to equal another train, the other train is odd. Also note that we can build an even number train with two rods both of the same color, here both green.

    Who can name the even rods? Sarah.
Sarah: Red, purple, dark green, brown, orange.
Mentor: So which rods represent odd numbers? Alexander.
Alexander: White, green, yellow, black, and blue.
Mentor: We will see these sequences again in many places.

Mentor: Class, did you learn something new?    Yes. What?
Mentor: Class, did you have some fun?    Yes. What did you like?

**Act 2    Scene 3    Black**

Mentor: Now we will consider black (seven), asking ourselves if it is prime or composite. As with yellow and dark green rods, we should justify our claims by showing the appropriate trains. How about black? Prime or composite? Sarah.
Sarah: I have found two one-color trains: white and black, making black composite.

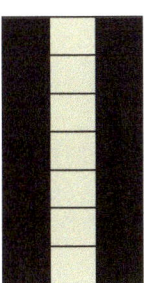

Mentor: Does everybody agree with Sarah?
Tom: Well, Sarah has two one-color trains equal to black, but that would only mean that black, or 7, is prime.
Mentor: Would you please explain that?
Tom: You said that a color, or number, is prime if it has just two one-color trains equal in length. Maybe there is another one-color train equal to black, making black a composite number. So we have to make sure we eliminate all the other possibilities.

Mentor: Excellent. Has anyone found another one-color train, besides black or white, equal to a black rod to make it composite? (No response.) Just because nobody has found such a train, can we claim that none exists?
Sarah: No. We might have missed one.
Mentor: So what do we have to do?
Sarah: We have to look at all the colors shorter than black, to make sure we did not miss a third one-color train.
Mentor: Excellent. Sarah, tell us about red. Could a red train equal a black rod? What do you have for evidence?
Sarah: Three red rods make a shorter train; four a longer. See! ?

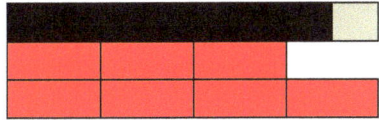

Mentor: Just so! Tom, could you show us that green cannot make a one-color train equal to black?

Tom: Easy! A train made of two green rods needs an additional white one to make it equal to the black train, but adding a third green overshoots the end of the black rod.

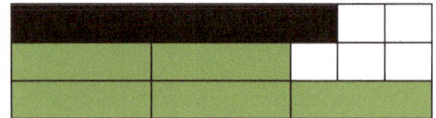

Alexander: May I do the purple trains?
Mentor: Yes.
Alexander: One purple rod comes up short, but two reach out beyond a black rod. Here is my evidence!

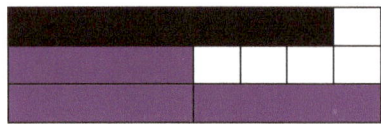

Mentor: That proves it! What else have we got?
Mary: Yellow and dark green.
Mentor: Mary, show us your argument against a possible yellow train equal to a black rod.

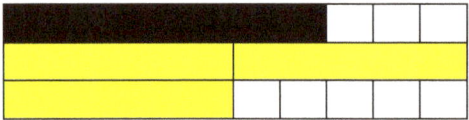

Mentor: Just right, Mary. Now Sarah would you show us that dark green does not do the job?
Sarah: Yes. Here.

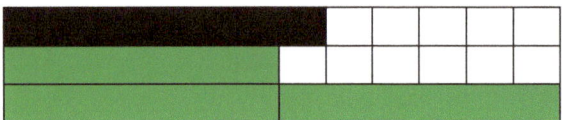

Mentor: After all this display, what can we now say?
Sarah: Seven is a prime number!
Mentor: We have completed our proof that black is a prime number.

**Act 2        Scene 4            Brown**

Mentor: Can anybody tell us about brown? Composite or Prime? Yes, Sarah.
Sarah: I can make a red train equal to a brown one, so brown is composite.

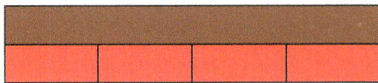

Mentor: Why can you claim that? Don't we have to find more than two one-color trains equal to the given train to show it is composite?
Sarah: Yes, but we know we can find brown trains and white trains equal a brown rod, so red makes the third one-color train. And so we have three! Also the number 8 is even.
Mentor: No argument there! We have demonstrated that 8 is a composite number.

Could anybody find yet another one-color train equal to a brown rod? George.
George: Two purple rods also make a train equal to the brown rod.

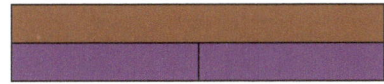

Mentor: Excellent. We now have a second way to show that brown is a composite number, but we only need one of them for our demonstration.

**Between the Scenes**
White, red, purple, and brown rods relate to each other in a unique way. Can you find it?
(Hint: How many white rods make a train as long as a red rod?)

**Act 2       Scene 5              Blue and Orange**

Mentor: We want to separate all the colors, except white, into prime and composite numbers. So help me with the color blue, Harry.
Harry: That's easy! Blue is composite, because an all-green train can equal a blue one.
Mentor: Show me!
Harry: Three green rods make a one-color train equal to the blue rod.

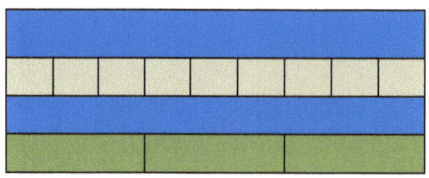

Mentor: Good! We know we can always find two colors, making trains equal to the given train: the given color and white. So Harry gave a correct answer.

To which category does orange belong? Alexander.
Alexander: Orange is prime.
Mary: No, it is composite!
Mentor: Mary, tell us how you know that orange is composite.
Mary: I can show you two different ways. Two yellow rods make a train equal to an orange rod. That could complete my argument, but I have another color: red, making orange even also. Here is my demonstration.

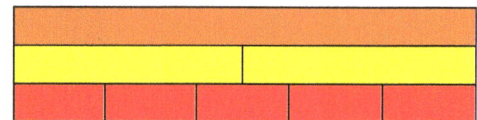

Mentor: Mary has shown us that two yellow rods make a train equal to an orange one, so orange belongs to the composite class. She could have given us red as well. Five red rods make a train equal to the orange rod also. And, of course, ten white rods makes a train equal to an orange train.
   At this point we have separated all the colors yellow and longer into prime and composite classes as follows:
   Prime: Y or 5, K or 7; and
   Composite: D or 6, N or 8, E or 9, O or 10.

Next time we will look at the shorter rods.

**Act 2        Scene 6        Red, Green, and Purple**

Mentor: Let us consider what we have learned about prime and composite numbers. Can anybody tell us when a color is prime? Sarah.
Sarah: A color is prime if there are exactly two one-color trains equal to it.
Mentor: Excellent! Is there any color that cannot have two different one-color trains equal to it?
Sarah: Yes. White can only have one one-color train equal to it: white.
Mentor: Fine. White is in a class by itself. White is not prime and white is not composite.
 So can somebody tell me about red? Tom.
Tom: Red is prime, because it has only two possible one-color trains equal to it: red and white.

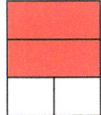

Mentor: Now we have prime numbers: R or 2, Y or 5, K or 7. Let's move on to green. Alexander.
Alexander: Green is also prime, because we cannot make a red train equal to a green rod. One red rod falls short by a white rod, but two red rods make a train too long by a white rod, making it also an odd number.

Mentor: Alexander has taken care of green. Now our list of prime numbers includes: R, G, Y, K or 2, 3, 5, 7, respectively. Who wants to show us about purple? Sarah.
Sarah: Purple is a composite number, because a red train with two rods equals a purple train. Here is my proof!

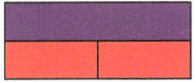

Mentor: Now our list of composite numbers includes: P, D, N, E, and O, or 4, 6, 8, 9, and 10 respectively.
Mentor: Class, did you learn something new?     Yes. What?
Mentor: Class, did you have some fun?     Yes. Why?

## Act 3      Counting

### Scene 1      Learn A Little, Gain A Lot

(If this scene seems too abstract, save it until after the class has learned to count by 2's, 4's, 6's, and 8's. We emphasize that the same or similar names are used in many counting situations.)

Mentor: Once we have learned to count to ten, one-by-one, we have also learned the basis for counting with some bigger numbers. Counting by hundreds from one hundred to one thousand follows the same pattern as counting to 10, except that we add on the word "hundred." We can read the numbers in the two "T's" below, first down the left-hand column and then down the right-hand column.

| by | 1's | by | 100's |
|----|-----|-----|-------|
| 1 | 6 | 100 | 600 |
| 2 | 7 | 200 | 700 |
| 3 | 8 | 300 | 800 |
| 4 | 9 | 400 | 900 |
| 5 | 10 | 500 | 1000 |

In words, we say: one hundred, two hundred, three hundred, four hundred, five hundred, six hundred, seven hundred, eight hundred, nine hundred, one thousand. We have the special name for ten hundred: one thousand. We do not have Cuisenaire rods of one hundred centimeters length, but a meter stick gives us the desired train. How many orange rods do we have to use to make a train as long as a meter stick? Tom.
Tom: I have used ten orange rods to make a train as long as a meter stick.
Mentor: Excellent. We will not actually try to make a white train as long as a meter stick, but if we did, how many white rods would we need?
Mary: One hundred.
Mentor: Exactly!
(The Mentor asks enough students to count to one thousand by hundreds to be sure everybody can do it easily.)

Mentor: This counting-by-hundreds pattern we have just learned repeats itself when we count by thousands, with two exceptions: change the word "hundred" to "thousand" and say "ten thousand" at the end. Unlike the counting by hundreds where we said "one thousand" rather than "ten hundred," we have no special name for ten thousand in English. So again, I want each of you to be able to count to ten thousand by thousands, just as you did counting to one thousand by hundreds.

|  by  | 1000's |
|-----:|-------:|
| 1000 | 6000   |
| 2000 | 7000   |
| 3000 | 8000   |
| 4000 | 9000   |
| 5000 | 10,000 |

Mentor: Notice we write a comma in the number 10,000 with three zeros on the right. Generally we put a comma between the third and fourth digits when the numbers exceed ten thousand.

(Again the Mentor has as many students count to ten thousand as necessary to know that all can count this way. The Mentor may mention that a meter has a thousand millimeters, leading to the ten-meter stick extending to ten thousand millimeters.)

Mentor: Counting by tens to one hundred requires a little more change in vocabulary, but not much. After we say "ten," we use the usual two, three, with "-ty" tacked onto the end. We say: ten, twenty, thirty, forty, fifty, sixty, seventy, eighty, ninety, one hundred. The exceptions concern twenty, instead of "two-ty"; fifty, instead of "fivety"; and thirty, instead of "threety."

| by | 10's |
|---:|-----:|
| 10 | 60   |
| 20 | 70   |
| 30 | 80   |
| 40 | 90   |
| 50 | 100  |

Mentor: We conclude these forms of counting by including the "teens." We will count to twenty by ones, starting with eleven, another special name, this time for the number coming after ten: 11. We observe a second exception with twelve, coming after eleven: 12. All the rest we can tack "teen" onto the number: thirteen, fourteen, fifteen, sixteen, seventeen, eighteen, nineteen, twenty, and make some slight changes to make the words sound better. We do not say "threeteen" or "fiveteen" but rather "thirteen" and "fifteen" respectively. Let us write the numbers in our T form and say them together.

| by | 1's |
|---:|----:|
| 11 | 16  |
| 12 | 17  |
| 13 | 18  |
| 14 | 19  |
| 15 | 20  |

Mentor: With the exception of 20, we create these numbers from our counting 1 to 10, by placing the digit '1' on the left side.

**Act 3          Scene 2          Counting by 2's To Twenty**

Mentor: Now that we know our colors pretty well and can count using our original list of numbers, we can move on to the next idea: counting by skipping over numbers in a regular way. How many children are there in this classroom? Let's start here, counting together: 1, 2, 3, …, 20. We have counted the students one at a time. Sometimes we can count objects two at a time, using only the even numbers, saving time and, perhaps, organizing our count. We want to see how to count by 2's now. Let's make two parallel trains: one red, the other white. We will count the white rods, but only say the number when our current white rod coincides with the end of a red rod. For example, we can count by 2 to six this way. Note that white rods have a grayish color to distinguish them from the paper.

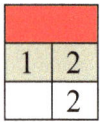

The first red rod equals a train of two white rods. Counting by 2's we would only say 2, the number in the third row.

|   |   |   |   |
|---|---|---|---|
| 1 | 2 | 3 | 4 |
|   | 2 |   | 4 |

Counting the number of white rods in the second picture by 2's we say: 2, 4, the two numbers in the third row.

|   |   |   |   |   |   |
|---|---|---|---|---|---|
| 1 | 2 | 3 | 4 | 5 | 6 |
|   | 2 |   | 4 |   | 6 |

Finally, we count up to six by two's, saying: 2, 4, 6, saying again the numbers in the third row.

(Continuing to build longer trains, we can count up to twenty, but we miss another pattern that will help students enjoy the power of recognition. So we draw T's, filling in the left side with the first half of the counting numbers, the right side with the second half.)

Mentor: I want you all to get out pencil and paper to record your counting in a way that will reveal a helpful pattern. Draw a large T, and above it write "by 2's" as shown below. This way. What do you notice about the **units** digits on both sides of the T? If you can count by one's to twenty, then you can count to 20 by 2's, simply not saying the odd numbers.

```
 by   2's
 ─────────
  2  | 12
  4  | 14
  6  | 16
  8  | 18
 10  | 20
```

Mentor: The **units** digits are the right-most digits in any number. Change the color of your pencil and put a box around the units digits. Your T should look something like this:

```
       by      2's
      ┌─┐    ┌─┐
      │2│  1 │2│
      │4│  1 │4│
      │6│  1 │6│
      │8│  1 │8│
    1 │0│  2 │0│
      └─┘    └─┘
```

Recording the units digits on the left side, we find: 2, 4, 6, 8, 0. On the right side they are: 2, 4, 6, 8, 0. What can you say about this sequence of digits? Sarah.

Sarah: They are the same!

Mentor: Just so. So if we can write the first half of our count, from 2 to 10, what do we have to do to write the second half?

Sarah: Just put a 1 in front of each digit, except the last, and change the 1 in 10 to a 2 to make 20.

Mentor: Wow. You've got it!

Making almost anything seem silly, or at least different, provides one way to remember it. A school cheer may not usually teach mathematics, but this one could provide an entertaining way to remember the sequence of even numbers.

### School Cheer
2, 4, 6, 8,
Who do we appreciate?
Evens! Evens! Evens!

In our next scene, we will consider counting by 4's and find a similar pattern, a pattern we should also find easy to remember.

**Between the Scenes** (Thanks to D. A. Penner)
Starting at one, count to nineteen by twos. Make a T to see the parallel.

**Act 3      Scene 3           Cuisenaire Color Coding**

Mentor: This lesson teaches us about the mixing of colors in something we call the "**color wheel**," often taught in art classes. The creators of the rods took care that certain rods would form a grouping whether we talk about length or color. So let's all take up our paintbrushes, paper and trays of colors to see what happens when we mix the different colors. Your trays have the three **primary colors**: red, yellow, and blue. We call them "primary" because we can start with them to create all, or almost all, the other colors. Make sure that your paintbrushes are always clean when you go to scoop out a new color, otherwise you will end up with no primary colors. You also only need a small amount of each color to explore this topic.

Mentor: Mary, tell us what color you get when you mix red and blue.
Mary: I got purple.
Mentor: Good, did everybody get the same result? (Walking around to check other students' results.) Now tell us, Mary, what color rod has the same length as a train made up of two red rods.
Mary: Purple.
Mentor: Have you perhaps discovered an accidental matching?
Mary: Hmmm. I don't know. You seem to be suggesting something.

Mentor: I want everybody to wash out their brushes and to try to mix just red and yellow. What color do you get?
Alexander: I get orange.
Mentor: Alexander, do you make any connection between the red and the yellow trains and the orange rods?
Alexander: Yes. Two yellow rods make a train as long as an orange rod, and five red rods make a train as long as an orange rod.
Mentor: Have you discovered something, perhaps a rule, about colors and lengths?
Alexander: We use red to make orange in color mixing, and we use red rods to create a train equal to the orange rod, and we use two yellow rods to make a yellow train equal to the orange rod!
Mentor: Excellent!

Mentor: Again, let's wash out our brushes, so we can mix all three primary colors. This experiment might not be quite so clear as the first two, but try to mix about equal parts of red, yellow and blue. What color do you get? Tom.
Tom: I get a muddy brownish color, and I know that four red rods make a train equal to a brown rod.
Mentor: Good observation. Does anybody else see another train possibility? Sarah.
Sarah: Two purple rods make a train equal to the brown rod, and purple has red in it!
Mentor: When we counted by twos and recorded our results in a T, what color did we find in each of the listed rods? Sarah.
Sarah: We found red rods or rods containing some red.
Mentor: Did you see any exceptions?
Sarah: Dark green does not suggest it contains any red.

Mentor: You have found, what feels to me, a weakness in the color-length system. I find it difficult to mix a dark green color from the primary colors without making a muddy, brownish color. However, we will consider that dark green has some red mixed into the yellow and blue to make the green darker. Perhaps some of you can come closer to dark green than I could.

Counting dark green as a red mixture along with the other colors we have looked at, what can we observe about red and the numbers we list when we count by two? Of course, you have all made the observation: Even numbers always have at least some red in their rods.

When we move on to the next scene or two, we will look at counting by the even numbers to look for patterns similar to those found in the by-2's T. Please make sure you return your brushes in clean condition.

**Between the Scenes**
1. Count by colors from white to orange.
2. Starting at red and skipping one color, count by colors to orange.
3. Starting at white and skipping one color, count by colors to blue.
4. Starting at green and skipping two colors, count by colors to blue.
5. Count backwards by colors from orange to white.

**Act 3**      **Scene 4**      **Counting By 4's To Forty**

Mentor: We will learn to count by 4's to forty just as we did learning to count by 2's to twenty. Of course, we will use purple rods instead of red ones, but also we have already seen a pattern we did not know before. We will work with the rods and our T's together. We start with 4, the number we find when we count the number of white rods needed to make a train equal to one purple rod. We say only the numbers in the fourth row, here 4.

|   |   |   |   |
|---|---|---|---|
| 1 | 2 | 3 | 4 |
|   |   |   | 4 |

Now we create our T with the first five numbers on the left, the second on the right. This will help us discover a pattern for 4's the way we discovered a pattern for 2's.

```
        By  4's
         4 |
           |
           |
           |
```

Mentor: Who can tell me what comes next?
Tom: 8.
Mentor: Can you show us why?
(Tom makes a train with two purple rods and another with eight white ones.)
Tom: I counted to 8 but only said 4 and 8.

| 1 | 2 | 3 | 4 | 5 | 6 | 7 | 8 |
|---|---|---|---|---|---|---|---|
|   |   |   | 4 |   |   |   | 8 |

Mentor: Excellent. Who knows what comes next this time? Sarah.

Sarah: 4, 8, 12. Here is my evidence!

Mentor: Just so. Sarah, show me your T.
(Sarah shows her T.)

```
By   4's
 4  |
 8  |
12  |
    |
```

Mentor: What comes after 12?
George: 15.
Mentor: Please, show us how you got to this answer. Where is your pair of trains?
(George quickly assembles the two trains.)

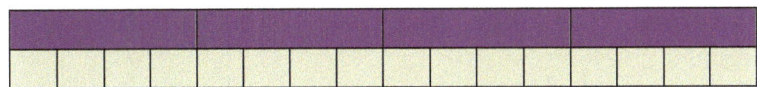

How many white blocks can you count in your second train?
George: 16.
Mentor: Can you make a red train for each purple rod?
George: Yes.
Mentor: So we could count by two's to reach 16. Did fifteen ever appear in our counting to 20 by 2's?
George: No.
Mentor: The numbers we use to count by 2's we call **even** numbers. If we add a red train to another red train, will we still have a red train?
George: Of course!
Mentor: So adding two even numbers gives us another even number.
   Let's continue to count by 4's. Mary, what follows 16?
Mary: 20. Here are my trains and my T with the first five numbers on the left!

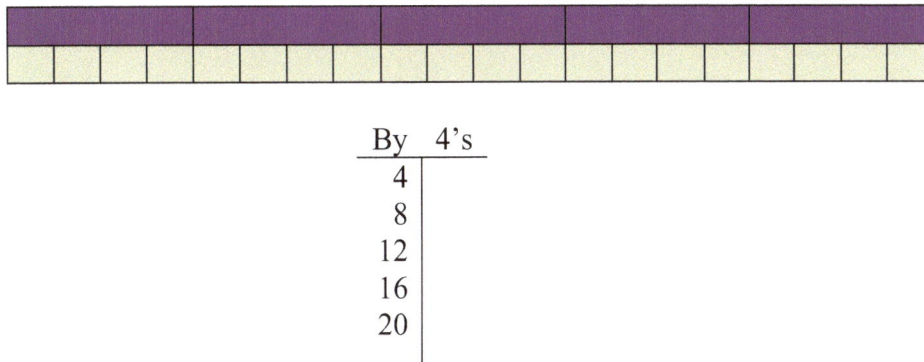

```
By   4's
 4  |
 8  |
12  |
16  |
20  |
    |
```

Mary: I have run out of white rods. How can I go on?
Mentor: What color rod would you use to replace 10 white ones?
Mary: Orange.

Mentor: Alexander, what comes after 20?
Alexander: 24, and here is my new pair of trains.

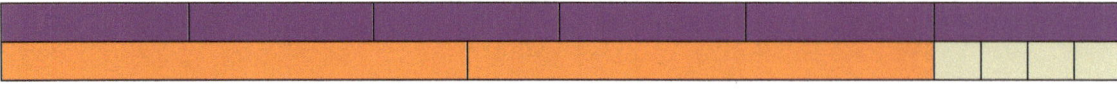

And here is my T!

|  By | 4's |
|---:|:---|
| 4 | 24 |
| 8 |    |
| 12 |    |
| 16 |    |
| 20 |    |

Mentor: Who can see the pattern forming on the right side of the T?
Sarah: 28 comes after 24.
Mentor: Excellent. What evidence can you offer? (Train below starts at 18.)

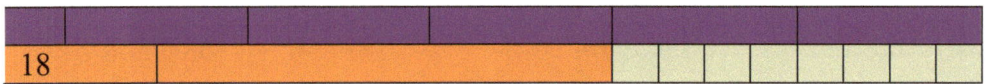

Sarah: And here is my T!

|  By | 4's |
|---:|:---|
| 4 | 24 |
| 8 | 28 |
| 12 |    |
| 16 |    |
| 20 |    |

Mentor: Exactly! Can anybody guess at the rest of the table?
Mary: I think the unit's digits are going to be the same on the right as on the left!
Mentor: Wow! So how do you fill in the rest?
Mary: 32, 36, and 40.
(Mary quickly shows her extended trains and her T.)

|   | By | 4's |   |
|---|----|----|---|
|   | 4  | 2  | 4 |
|   | 8  | 2  | 8 |
| 1 | 2  | 3  | 2 |
| 1 | 6  | 3  | 6 |
| 2 | 0  | 4  | 0 |

Mentor: Now we know how to count by 4's to 40. How did we do this?
Alexander: We started with 4 (or one purple rod) and added 4 until we had five numbers in the left-hand column, reaching 20 or half way to 40. Then we record the second group of five numbers on the right side of the T.

We just keep adding another purple rod until we could count 40 white ones. Don't forget to put red boxes around the unit's digits!

Mentor: We just had to add purple rods. So now we go on to counting by 6's.

**Between the Scenes**
1. Starting at one, count by fours to 37.
2. Starting at two, count by fours to 38.
3. Starting at three, count by fours to 39.
4. Count backwards by fours from 40 to 4.

**Act 3**     **Scene 5**     **Counting By 6's To Sixty**

(This scene may call for the Mentor taking some students through addition iterations, such as 6 + 6, 16 + 6, . . ., 86 + 6, until the pattern becomes obvious.)

Mentor: We can now count by 6's to 60, recognizing the method used for counting by 2's and by 4's. We can also remember that each dark green rod equals three red rods, so we know that we expect only even numbers in our T-tables. Who would like to start us off? Jane.

Jane: 6, 12, 18, 23.

Mentor: Show us your train with 4 dark green rods and a parallel train of white rods.

[train diagram: 4 dark green rods with parallel white rods labeled 6, 12, 18, 23, 24]

Jane: Oh, I see. We have twenty-four white rods, not 23!

Mentor: And 23 is odd, not even. Yet the train consists of just even numbers, namely dark green rods.

```
By  6's
 6 |
12 |
18 |
24 |
```

Mentor: What number comes after 24?

Alice: 28.

Mentor: Alice, what number do you get when you add a purple rod and a dark green one? Or adding 4 and 6?

Alice: (Puzzled?) I am adding a dark green rod to 24!

Mentor: You have not seen a pattern, a pattern making this calculation trivial. Start with just a purple rod and a dark green one. What rod equals that train?

Alice: Orange. 10.

Mentor: Excellent. Now make a train of an orange and a purple rod, adding a dark green one. What do you get now?

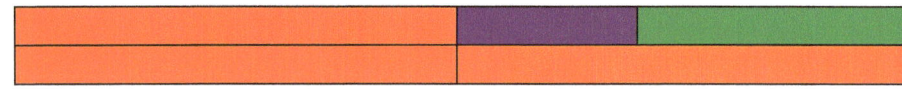

Alice: 20.

Mentor: How do you show that result?

Alice: I make a parallel train with two orange rods and find that the second train equals the first.

Mentor: Now add a dark green rod to a train made of two orange rods and one purple one. What will you have in the equal parallel train?

Alice: 30, or three orange rods.
Mentor: Just to work on this pattern a little further, Alice, what number will we reach if we add one dark green rod to a train of three orange rods and one purple one?
Alice: 40.
Mentor: To a train of four orange rods and one purple?
Alice: 50.
Mentor: Five orange rods and one purple?
Alice: 60.
Mentor: Six?
Alice: 70.
Mentor: Does everybody see the pattern?
We continue counting by sixes.

| By | 6's |
|---|---|
| 6 | |
| 12 | |
| 18 | |
| 24 | |
| 30 | |

Mary: Can I fill in the rest?
Mentor: If you like.

| By | 6's |
|---|---|
| 6 | 36 |
| 12 | 42 |
| 18 | 48 |
| 24 | 54 |
| 30 | 60 |

Mentor: Class, did Mary complete the table correctly?
Class: Yes!
Mentor: What do we do now?

Class: Draw a different color box around the units digits.

|   | By | 6's |   |   |
|---|----|----|---|---|
|   | 6  | 3  | 6 |
| 1 | 2  | 4  | 2 |
| 1 | 8  | 4  | 8 |
| 2 | 4  | 5  | 4 |
| 3 | 0  | 6  | 0 |

Mentor: What pattern do we see being repeated?
Class: The units digits: 6, 2, 8, 4, 0.
Mentor: It would appear we can all count by sixes! Next comes counting by 8's.

**Between the Scenes**
1. Count by sixes from 1 to 55, recording your count in a T.
2. Vary the count's starting position, helping students see the patterns, such as 7 + 6 = 13 turning into 37 + 6 = 43 several steps later.
3. Count backwards from 60 to 6 by 6's.

**Act 3**  **Scene 6**  **Counting by 8's**

(This scene and the counting by 6's can flow more easily if there are some rods of length 20 and 30 cm. Few sets of rods have enough orange rods to show all the calculations.)

Mentor: Who would like to get us started counting by 8's? Frank.
Frank: 8, 16, 24, 30.
Mentor: Frank, please show us how you arrived at these numbers. In particular, show us your result when you add a brown rod to a purple rod. What color rod do you have to add to an orange rod to make the parallel train of equal length?
Frank: A purple plus a brown rod equals a train of orange and red, see?

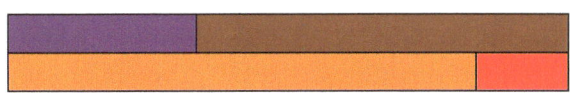

Mentor: So 4 + 8 = 12. What about 14 + 8? (Writing on the board: 4 + 8 = 12, etc.)
Frank: 14 + 8 = 22.
Mentor: 24 + 8?
Frank: 24 + 8 = 32.
Mentor: Excellent! Try 34 + 8!    Frank: 42.    Mentor: 44 + 8?    Frank: 52.
Mentor: 54 + 8?    Frank: 62!    Mentor: 64 + 8?    Frank: 72.
Mentor: 74 + 8?    Frank: 82.    Mentor: 84 + 8?    Frank: 92!
Mentor: 94 + 8?    Frank: 102. (This usually ends up a rapid fire back and forth.)
Mentor: Excellent. I think you have seen the pattern. Let's fill in the T. What comes after 24?
Frank: 32.

```
     By  8's
     ────────
      8 │
     16 │
     24 │
     32 │
        │
```

Mentor: Frank, can you finish off the left side of the T?
Frank: Yes. After 32 comes 40.
Mentor: Correct. Now let's complete the right side of the T. George.
George: 48, 56, 64, 72, 80.
Mentor: That fits the pattern. Write the results in your T.

George writes:

|  By  | 8's |
|------|-----|
|  8   | 48  |
| 16   | 56  |
| 24   | 64  |
| 32   | 72  |
| 40   | 80  |

Mentor: Now make a box around the units digits on both sides of the T. Sarah, read the digits in your boxes.

|   | By |   | 8's |
|---|----|---|-----|
|   | 8  | 4 | 8   |
| 1 | 6  | 5 | 6   |
| 2 | 4  | 6 | 4   |
| 3 | 2  | 7 | 2   |
| 4 | 0  | 8 | 0   |

Sarah: 8, 6, 4, 2, 0 and 8, 6, 4, 2, 0 again!
Mentor: What pattern do you find in this sequence of unit's digits?
Sarah: We are just counting backwards from 8 by 2's.
Mentor: Why do we see the units digits go down 2 each step? What would happen if we added 10 instead of 8?
Sarah: If we added 10 each time the unit's digits would always be the same. Perhaps we could think of adding 10 to get the same digit and then subtracting 2.
Mentor: So adding 10 and subtracting 2 gives us the same result as adding 8. How about the ten's digits?
Sarah: They just count up, except for 40 and 48 repeating the 4.
Mentor: Excellent.

**Between the Scenes**
1. Starting at one, count by 8's to 73.
2. Starting at two, count by 8's to 74.
3. Count backwards from 80 by 8's.

**Act 3**      **Scene 7**      **Counting by 3's, An Odd Job**

Mentor: When we counted by an even number, say 6, what did we find for units digits?
Alexander: Even digits!
Mentor: Namely?
Alexander: 0, 2, 4, 6, 8.
Mentor: Just so! When we count by 3's we will find a different pattern. So look for it!
Jane, try counting by 3's, and building the parallel trains to show your results.

Jane: The first part is easy: 3, 6, 9, 12. My parallel trains show it.

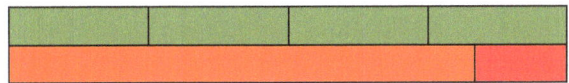

Mentor: Fine! What comes next?
Jane: 15.
Mentor: Do you find just odd numbers in this series?
Jane: No. They alternate odds and evens.
Mentor: So let us fill the left side of our T, counting by 3's.

```
         By   3's
          3  |
          6  |
          9  |
         12  |
         15  |
```

Mentor: When we added two 3's, two odd numbers, we got a 6, an even number. When we added 6, an even number, and 3, an odd number, we got 9, an odd number. Think about these results as we complete the T. George, complete the right-hand side, showing us your parallel trains to start off.

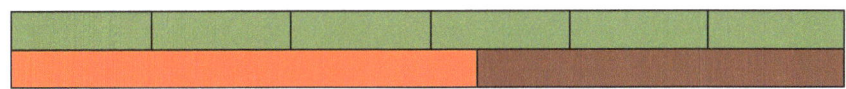

George: I have six green rods in one train and a parallel train with an orange and a brown rod. That makes 18.
Mentor: Can you complete the right side of the T?
George: Here!

|  By | 3's |
|----:|----:|
|  3  | 18  |
|  6  | 21  |
|  9  | 24  |
| 12  | 27  |
| 15  | 30  |

Mentor: Excellent! What do you notice about the unit's digits in the two columns?
George: They do not equal each other the way the even numbers did.
Mentor: How much do you have to add to 3 to get 18? To 6 to get 21? To 9 to get 24? Have you found another pattern?
George: I always have to add 15 to the left-hand number, or just half of 30, to get the number in the right column!
Mentor: Notice that in Act 3 Scene 1 we started the pattern of counting half way to 10 on the left side of our T, and then we counted the rest of the way on the right side. Looking at that first T, repeated below, we see that 6 was to the right of 1, 7 to the right of 2, and so on. Each number on the right is 5 more than its partner on the left, and the difference equals half of the last number, here half of 10.

|  By | 1's |
|----:|----:|
|  1  |  6  |
|  2  |  7  |
|  3  |  8  |
|  4  |  9  |
|  5  | 10  |

Mentor: Let's review what happens when we add odd numbers. What do we see happening each time we add 3 to 3, or 3 to 9? What kind of number do we get?
George: We get an even number. The two white rods in the sum we can replace with a red rod.
Mentor: What outcome do we find adding an odd and an even number?
George: The combined train will have just one white rod, so we obtain an odd number!
Mentor: Do you think your argument only works for adding 3's?
George: No. It will always work.

Mentor: Class, did you learn something new?   Yes.
Mentor: Class, did you have some fun?   Yes.

**Between the Scenes**
1. Starting at one, count to 28 by 3's.
2. Starting at two, count by threes to 29.
3. Count backwards by threes from 30.

**Act 3        Scene 8        Count By 5's**

Mentor: Most of you probably know how to count by 5's already, but we should take a look anyway. We picture the colors yellow-orange-yellow-orange and so on. The rhythm we fall into counting by fives helps make the sequence easy and fun. Can anyone tell us how to count by fives to fifty?
George: 5, 10, 15, 20, 25, 30, 35, 40, 45, 50.
Mentor: Fine. Now let us write the numbers in a T, thinking about the odd and even numbers. What patterns do you see?

| By | 5's |
|---|---|
| 5 | 30 |
| 10 | 35 |
| 15 | 40 |
| 20 | 45 |
| 25 | 50 |

George: I see the same pattern as we had with counting by 3's, namely odd, even, odd, even. Also, I see that the last number on the left is half the last number on the right.
Mentor: What pattern do you see in the odd numbers?
George: The odd numbers all end with 5 and the even with 0.
Mentor: Why should the sum of two odd numbers be an even number? Alexander.
Alexander: Each odd number is a red train with one white rod, so when we put them together we have a red train with two white rods. We can replace the two white rods with a red one and have a one-color red train. We did the same thing with sums of three.
Mentor: Wow. Good explanation! Can someone else explain why the sum of an odd number and an even number gives us another odd number? Sarah.
Sarah: We can make any even number with a one-color red train, the odd number with a one-color red train and a single white rod. Putting these together, we have a one-color red train with just one white rod, making it an odd number.
Mentor: Well done! We have seen three "rules" about adding odd and even numbers. Can anybody tell us what the rules say? Tom.
Tom: <u>The sum of two even numbers and of two odd numbers is even</u>, while <u>the sum of an odd and an even number is odd</u>.
Mentor: There they are! These so-called rules give us a guide to check our arithmetic. If you are adding two odd numbers and get an odd number, you know you have made some kind of mistake somewhere. Go back and look at what you have done.
Can anybody see another pattern, comparing the two sides of the T?
Jane: Yes. If I add 25 to the number of the left, I will get the number on the right.
Mentor: Excellent.

**Act 3        Scene 9              Surviving The Sevens**

Mentor: As we already have discovered in Act 2 Scene 3, the black rod, 7, is a prime number. Now we want to count by sevens to 70, detecting any patterns as we move along. Thinking back on our recent counting scenes, does anybody anticipate a pattern? Mary!
Mary: We will see the odd-even-odd-even pattern?
Mentor: What gave you that idea?
Mary: It happened with 3 and 5.
Mentor: Good. Do you want to start us off?
Mary: 7, 14, 21, 28, 33.

Mentor (Writing 8 + 7 = ): Mary, Calculate the sum of 8 and 7, and show us your calculation with the rods.
Mary: 8 + 7 = 15.

Mentor: Yes, we add an even and an odd number, so the result is odd. Now add eighteen and 7. What do you do to your two trains to show your calculations?
Mary: All I do is put two orange rods on the left of the two trains I already have. So 18 + 7 = 25.

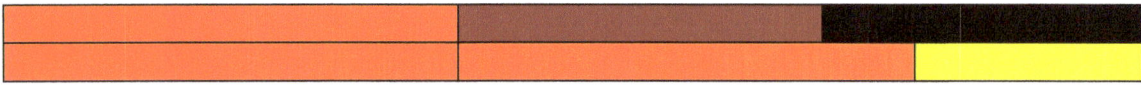

Mentor: Now 28 + 7?        Mary: 35        Mentor: 38 + 7?        Mary: 45.
Mentor: 48 + 7?        Mary: 55.
Mentor: What mistake did you make?
Mary: 28 + 7 = 35, not 33.
Mentor: I think you have the pattern now. Let's draw our T's, filling in the left side.

By 7's
| 7 |
| 14 |
| 21 |
| 28 |
| 35 |

Mentor: George, can you direct us through the right side?
George: Sure: 42, 49, 56, 63, 70.
Mentor: Everyone should now have the 7's T filled, as George dictated.

|  By  | 7's |
|---|---|
| 7 | 42 |
| 14 | 49 |
| 21 | 56 |
| 28 | 63 |
| 35 | 70 |

Mentor: What pattern relates the left and the right sides of the T?
George: The right side equals the sum of the left side and 35, or half of 70.
Mentor: Terrific! If nobody has any questions, we can go on to counting by 9's.

Between the Scenes
1. Starting at one, count by fives to 46.
2. Starting at two, count by fives to 47.
3. Starting at one, count by sevens to 64.
4. Starting at two, count by sevens to 65.
5. Count backwards from 50 by fives.
6. Count backwards from 70 by sevens.

**Act 3**  **Scene 10**  **Counting By Nines**

Mentor: In this scene, we will learn to count by 9's. We could proceed in the same fashion as we have with our other counting numbers, but I want to introduce a special property of numbers that can always be equal to a one-color blue train. You should find several familiar patterns as we go. You may, of course, make a long train of blue rods with a parallel train to give the same results.

| #9's | 1 less | + to get 9 | # |
|---|---|---|---|
| 1 | 0 | 9 | 09 |

We will make four columns, asking standard questions. In the first column, we will count from 1 to 10. In the second, we will record the number one less than that in the first column. In the third, the number we must add to the number in the second column so the sum equals 9. Finally, in the fourth, we record our count, placing the number in column two in front of the number in column three. While this may seem complicated at first, it soon becomes simple. Tom, help me fill in the first row. We start with 1.
Tom: 1 in the first.
Mentor: Good. What number is one less than 1?
Tom: 0 in the second.
Mentor: What number do you add to 0 to get 9?
Tom: 9.
Mentor: Yes, 9 in the third column. Making a two-digit number out of the digits in the second and third column, what do you get?
Tom: 09 in the fourth.
Mentor: Excellent. We do not usually write 09 for nine, just 9, but here we look for a pattern and the 0 before the 9 will help make this pattern clear. Sarah, would you help us fill the second row of our table? Counting one at a time, what comes after 1?
Sarah: 2
Mentor: What rod is one less than 2?
Sarah: 1
Mentor: 1 + what number equals 9?
Sarah: 8
Mentor: Excellent. Writing these two digits next to each other, gives us?
Sarah: 18

| #9's | 1 less | + to get 9 | # |
|---|---|---|---|
| 1 | 0 | 9 | 09 |
| 2 | 1 | 8 | 18 |

Mentor: If anyone builds a train of two blue rods, what will the parallel and equal train consist of?
Alexander: An orange rod and a brown one, making eighteen!


Mentor: Now we will complete the table. Jane, what should we write in the next row?
Jane: 3, 2, 8, 28
Mentor: What must we add to 2 to get 9? Does 2 + 8 = 9?
Jane: No. 2 + 7 = 9
Mentor: Exactly! So, now complete the row.
Jane: 3, 2, 7, and 27.
Mentor: You have it.
(Mentor calls on enough students to get the table filled.)

| #9's | 1 less | + to get 9 | # |
|------|--------|------------|-----|
| 1 | 0 | 9 | 09 |
| 2 | 1 | 8 | 18 |
| 3 | 2 | 7 | 27 |
| 4 | 3 | 6 | 36 |
| 5 | 4 | 5 | 45 |
| 6 | 5 | 4 | 54 |
| 7 | 6 | 3 | 63 |
| 8 | 7 | 2 | 72 |
| 9 | 8 | 1 | 81 |

Mentor: Tom, what can we say about the sum of the two digits of a one-color blue train?
Tom: The sum equals 9.
Mentor: Mary, what pattern do you see in the numbers in the first column?
Mary: They count by ones from 1 to 9.
Mentor: Sarah, what pattern do you see in the numbers in the second column?
Sarah: They count from zero to 8 by ones.
Mentor: George, what pattern do you see in the numbers in the third column?
George: The numbers count backwards from nine to one.
Mentor: Jane, do you have trouble counting by 9's now?
Jane: No! I just count up with the tens digit and down with the units digit. 09, 18, 27, 36, 45, 54, 63, 72, 81, 90. It's easy now!

Mentor: You have learned an important pattern for 9's. We will talk about it more later.

**Act 3**          **Scene 11**          **Collecting Counting**

Mentor: Today we will collect our counting into one big table, a table you will find useful in the future. We will use graph paper for our table, because neatness will help make the patterns more visible. Follow my directions carefully!

| 1 | 2 | 3 | 4 | 5 |   |   |
|---|---|---|---|---|---|---|
| 2 |   |   |   |   |   |   |
| 3 |   |   |   |   |   |   |
| 4 |   |   |   |   |   |   |
| 5 |   |   |   | X |   |   |
|   |   |   |   |   |   |   |
|   |   |   |   |   |   |   |

    From the upper left corner, count 5 cells across and from the same corner cell 5 down. Note that we call the horizontal lines of cells **rows** and the vertical lines **columns**. X marks the cell where our table starts. To keep track of the row and column we are looking at we will enter the numbers from 1 to 12 on the third row, starting at column 5 and skipping every other cell.

The table below shows the arrangement.

| 1 | 2 | 3 | 4 | 5 |   |   |   |   |   |   |   |   |   |   |   |   |   |   |   |   |
|---|---|---|---|---|---|---|---|---|---|---|---|---|---|---|---|---|---|---|---|---|
| 2 |   |   |   |   |   |   |   |   |   |   |   |   |   |   |   |   |   |   |   |   |
| 3 |   |   |   | 1 |   | 2 |   | 3 |   | 4 |   | 5 |   | 6 |   | 7 |   | 8 |   | 9 |
| 4 |   |   |   |   |   |   |   |   |   |   |   |   |   |   |   |   |   |   |   |   |
| 5 |   |   | X |   |   |   |   |   |   |   |   |   |   |   |   |   |   |   |   |   |
|   |   |   |   |   |   |   |   |   |   |   |   |   |   |   |   |   |   |   |   |   |
|   |   |   |   |   |   |   |   |   |   |   |   |   |   |   |   |   |   |   |   |   |

(third row continues: 10, 11, 12)

|   |   |   | 5 |   |   |   |   |   |   |   |   |   |   |   |
|---|---|---|---|---|---|---|---|---|---|---|---|---|---|---|
|   |   |   | 1 | 2 | 3 | 4 | 5 | 6 | 7 | 8 | 9 | 10 | 11 | 12 |
| 5 | 1 | X |   |   |   |   |   |   |   |   |   |   |   |   |
|   | 2 |   |   |   |   |   |   |   |   |   |   |   |   |   |
|   | 3 |   |   |   |   |   |   |   |   |   |   |   |   |   |
|   | 4 |   |   |   |   |   |   |   |   |   |   |   |   |   |
|   | 5 |   |   |   |   |   |   |   |   |   |   |   |   |   |
|   | 6 |   |   |   |   |   |   |   |   |   |   |   |   |   |
|   | 7 |   |   |   |   |   |   |   |   |   |   |   |   |   |
|   | 8 |   |   |   |   |   |   |   |   |   |   |   |   |   |
|   | 9 |   |   |   |   |   |   |   |   |   |   |   |   |   |
|   | 10 |   |   |   |   |   |   |   |   |   |   |   |   |   |
|   | 11 |   |   |   |   |   |   |   |   |   |   |   |   |   |
|   | 12 |   |   |   |   |   |   |   |   |   |   |   |   |   |

Now starting at row 5 column 3 count down by 1's just as you counted across on row 3, skipping a space between numbers. After you have filled in these two lists of numbers, your table is ready for counting skills.

Mentor: George, starting with the X cell, count by 1's to 12, entering your numbers below the corresponding numbers in row 3.
George: 1, 2, 3, 4, 5, 6, 7, 8, 9, 10, 11, 12.
Mentor: Alexander, count by 2's and enter the numbers on the row opposite the 2 (on the column 3) starting the entry under 1 on the top index row.
Alexander: 2, 4, 6, 8, 10, 12, 14, 16, 18, 20, 22, 24.

|    | 1  | 2  | 3  | 4  | 5  | 6  | 7  | 8  | 9   | 10  | 11  | 12  |
|----|----|----|----|----|----|----|----|----|-----|-----|-----|-----|
| 1  | 1  | 2  | 3  | 4  | 5  | 6  | 7  | 8  | 9   | 10  | 11  | 12  |
| 2  | 2  | 4  | 6  | 8  | 10 | 12 | 14 | 16 | 18  | 20  | 22  | 24  |
| 3  | 3  | 6  | 9  | 12 | 15 | 18 | 21 | 24 | 27  | 30  | 33  | 36  |
| 4  | 4  | 8  | 12 | 16 | 20 | 24 | 28 | 32 | 36  | 40  | 44  | 48  |
| 5  | 5  | 10 | 15 | 20 | 25 | 30 | 35 | 40 | 45  | 50  | 55  | 60  |
| 6  | 6  | 12 | 18 | 24 | 30 | 36 | 42 | 48 | 54  | 60  | 66  | 72  |
| 7  | 7  | 14 | 21 | 28 | 35 | 42 | 49 | 56 | 63  | 70  | 77  | 84  |
| 8  | 8  | 16 | 24 | 32 | 40 | 48 | 56 | 64 | 72  | 80  | 88  | 96  |
| 9  | 9  | 18 | 27 | 36 | 45 | 54 | 63 | 72 | 81  | 90  | 99  | 108 |
| 10 | 10 | 20 | 30 | 40 | 50 | 60 | 70 | 80 | 90  | 100 | 110 | 120 |
| 11 | 11 | 22 | 33 | 44 | 55 | 66 | 77 | 88 | 99  | 110 | 121 | 132 |
| 12 | 12 | 24 | 36 | 48 | 60 | 72 | 84 | 96 | 108 | 120 | 132 | 144 |

(The mentor calls on enough students to get the above table filled correctly.)

Mentor: Does anybody know what we have just constructed? Alexander.

Alexander: We have the multiplication table!

Mentor: So, with all the counting we have learned to do, we now find we have learned the multiplication tables. To become fluent with multiplication tables we will have to use them; there are many times we will want to multiply numbers. For the moment, we want to look at this table for familiar and new patterns. What do you observe about the first row and first column in our table?

Sarah: They equal each other. Just one reads across; the other, down.

Tom: That works for all rows!

Mentor: Would you explain that?

Tom: Sure. The eighth row and eighth column have the same entries. So do the other rows and columns with the same index number.

Mentor: What do we see where these row-column pairs intersect?

Jane: Don't we call those numbers' squares?

Mentor: Excellent. In our next scene, we will make a table of squares.

**Act 3**  **Scene 12**  **Identifying Factors**

Mentor: Having created your own multiplication tables by counting, we can now see some ways you can apply this knowledge. You know that $2 \times 3 = 6$, or alternatively that two green rods make a train as long as a dark green rod. Here we add two new words: factor and divisor. Another way of looking at $2 \times 3 = 6$ consists of seeing that $6/2 = 3$. We say 2 is a **factor** or **divisor** of 6, because I can find a number, 3, that multiplied times 2 equals 6. When I observe that I can make a red train the same length as a dark green rod, I am asserting the same thing: two is a factor of six. You have seen many patterns so far. Can anyone tell me how we know if 2 is a factor, or divisor, of any number?
Jane: When we counted by two's to twenty, we saw in our T's that the last digits were always: 2, 4, 6, 8, 0. These digits get repeated forever after.
Mentor: Just so. If I asked you about the number 56,934, could you tell me if two divides this number, or not, just by looking at it? If so, what would you look at? What would you do with the rest of the number?
Jane: I would look at the last digit, the 4. It belongs to the list I gave, so the whole number is even. I would ignore the rest of the number.
Mentor: Jane has shown us how to apply the **rule** for determining if a number is divisible by two: the unit's digit must be 0, 2, 4, 6, or 8. Can anybody else think of another rule, indicating that a certain number divides another number? Maria.

Maria: Sure. If the last digit of a number is a zero, then ten divides the number. 10 divides 70 and 10 divides 2,478,190.
Mentor: How did you come to this observation?
Maria: Counting by 10's always gave us a number ending in zero.
Mentor: That gives us a good start. Anybody else want to make an observation about divisor detection? Tom.
Tom: The rule for five is almost the same as for ten: the last digit must be a five or a zero.
Mentor: Terrific. Another?
Alexander: Yes. One divides every number, because one times any number equals that number.
Mentor: Very tricky. Yes. One is a factor of every number, but why do you think so, Alexander?
Alexander: We are able to make a white train the same length as any other train.
Mentor: Can somebody else think of another part of our multiplication tables, showing another rule?
Sarah: 9. We know that 9 divides another number if the sum of the digits of that other number equals 9.
Mentor: Your observation deserves more attention. We saw how the sum of the two-digit number multiple of 9 added up to 9, but what about three digit numbers? What about 972?


Sarah: The last two digits add up to 9.

Mentor: So we do not have to look at the hundred's digit? In the case of 972, we see that 9 times 100 equals 900, and 9 times 8 equals 72, so there is a blue train equal to 972. But try 172. The last two digits still sum to 9, but does 9 divide 172?

Sarah: No. Nine does not divide 172, because 9 times 16 (twice 8) equals 144, and 172 – 144 equals 28. We know that 9 does not divide 28. So we can make a blue train as long a 144, but not as long as 28.

Mentor: Excellent. What about 972 again?

Sarah: The sum of the digits equals 18, and 18 is a multiple of nine. With 172, we have the sum of the digits is 10, and 10 is not a multiple of 9. Nine is a factor of 972 but not of 172.

Mentor: Dividing by nine came easily because we learned to count by nines, noting the patterns: units digits count up $(1, 2, 3, \ldots, 9)$, tens count down $(9, 8, 7, \ldots, 0)$. We saw that the sum of the two digits always came out equal to 9. We simply expanded on that idea in recognizing that larger multiples of nine also had the sum of their digits equal to nine, or at least to a multiple of nine. We now look to see if we can get the same result for less effort.

If we have a long blue train, i.e. a multiple of nine train, and we took away some of the blue rods, what color train would we have left?

Jane: That's pretty obvious! We would still have a blue train.

Mentor: Exactly. So if I asked you to determine if the number 7182 is a blue train, what would you do?

Jane: I would add up $7 + 1 + 8 + 2$ to get 18 and observe that 9 times 2 equals 18. Then we know that 7182 is a multiple of nine.

Mentor: Just so. The question you asked yourself was if the sum of those digits formed a blue train. Suppose we see that $1 + 8 = 9$ and "cast away" that blue number, would the rest still be a blue train?

Jane: Yes. $7 + 2 = 9$.

Mentor: So we did not have to add the 1 and the 8 to the rest, just cast it away and look at the $7 + 2$. However, we could again recognize that $7 + 2$ equals 9, so we can cast that away also. This leaves us with zero. Is zero a multiple of 9?

Jane: No. 0 is too small to be a multiple of 9.

Mentor: What happens if you add up nine 0's? How much do you get?

Jane: 0.

Mentor: So, how much is 9 times 0?

Jane: 0.

Mentor: Then haven't you observed that 0 is a multiple of 9, namely $0 \times 9 = 0$?

Mentor: Great progress! Let's make a table of our findings as far as we have gone. The second column contains the condition for divisibility by the number in the first.

| 1 | 1 is a divisor of every number |
|---|---|
| 2 | 2 divides every number ending in: 0, 2, 4, 6, 8 |
| 3 | TBA (To Be Added) |
| 4 | TBA |
| 5 | 5 divides every number ending with 5 or 0 |
| 6 | TBA |
| 7 | TBA |
| 8 | TBA |
| 9 | 9 divides a number if the sum of its digits is a multiple of 9 |
| 10 | 10 divides a number if it ends with a 0. |
| 11 | TBA |
| 12 | TBA |

Mentor: Let's look now at 4. How do we know we can divide a number by 4, without dividing by 4? George.

George: If a number is even it is divisible by 4.

Mentor: What about 10? Ten is even, but is it divisible by 4? Can we make a purple train the same length as an orange rod?

George: Something is wrong.

Mentor: How about turning the **if-then**, or **conditional**, statement around. If four divides a number, is the number even?

George: Yes.

Mentor: Why?

George: If four divides a number, then I can make a purple train equal to that number's train, and I can replace each purple rod by two red rods. But I cannot make a purple train the same length as every red train.

Mentor: Let's shift to money for divisibility by 4. How many quarters in a dollar?

George: Four.

Mentor: How many fours are in 100?

George: 25.

Mentor: So, does 4 divide 1900?

George: Yes, because we can replace every hundred by 25 fours. We can make a purple train as long as a 1900 train.

Mentor: Now look at 1912. Is 4 a factor of 1912? Why? Joyce.

Joyce: Yes. We can make a purple train the length of 1900 and of 12. So adding these together, we get a purple train. The sum of two purple trains equal a purple train.

Mentor: Do you want to venture a guess at the rule for divisibility by four?

Joyce: <u>If 4 divides the number formed by the last two digits, then the number is divisible by 4</u>.

Mentor: We have that one now. How do we know if a number has 8 as a factor? (A hint: 2 = 2, 2 × 2 = 4, 2 × 2 × 2 = 8.) For a factor of 2, we look at the last one digit; for a factor of 4 we look at the last two digits; so for a factor of 8 we will look at the last how many digits?

Joyce: I don't know why, but I guess three digits.

Mentor: How many times 125 equals 1000? Joyce.

Joyce: 8 × 125 = 1000. So however many thousands we may have they will be divisible by 8, meaning that we only have to check the last three digits.

Mentor: Excellent. We only have to look at the number formed by the last three digits. Now we can update our table in two lines: 4 and 8.

| 1 | 1 is a divisor of every number |
|---|---|
| 2 | 2 divides every number ending in: 0, 2, 4, 6, 8 |
| 3 | TBA |
| 4 | 4 divides the number formed by the last two digits |
| 5 | 5 divides every number ending with 5 or 0 |
| 6 | TBA |
| 7 | TBA |
| 8 | 8 divides the number formed by the last three digits |
| 9 | 9 divides a number if the sum of its digits is a multiple of 9 |
| 10 | 10 divides a number if it ends with a 0. |
| 11 | TBA |
| 12 | TBA |

Mentor: Having seen how 2, 4 and 8 are related, what do you think 3 might be related to? Alexander.

Alexander: Three squared equals nine. So the rule for three will be related to the rule for nine.

Mentor: Excellent guess. Does three divide nine? Does nine divide 3?

Alexander: Green trains can always replace blue ones, but not the other way around. So three divides nine, but not the other way around.

Mentor: If I told you the rule for three resembled the rule for nine, what might you guess?

Alexander: If the sum of the digits is divisible by three, then the number is divisible by 3.

Mentor: Nice shot! Not everybody gets there so quickly. Let's go on to 6. What rule tells us that we can divide a number by 6? Sue.

Sue: The number has to be even, because 6 is even, and has to be a multiple of three, because 6 is a multiple of three.

Mentor: We can look at the justifications for the rules for six and twelve later, but we have almost filled our table.

| 1 | 1 is a divisor of every number |
| --- | --- |
| 2 | 2 divides every number ending in: 0, 2, 4, 6, 8 |
| 3 | 3 divides the sum of the digits of the number |
| 4 | 4 divides the number formed by the last two digits |
| 5 | 5 divides every number ending with 5 or 0 |
| 6 | 2 and 3 both divide the number |
| 7 | TBA |
| 8 | 8 divides the number formed by the last three digits |
| 9 | 9 divides a number if the sum of its digits is a multiple of 9 |
| 10 | 10 divides a number if it ends with a 0. |
| 11 | TBA |
| 12 | TBA |

Mentor: Looking at the multiples of eleven, can you see something that seems to remain constantly? Sue.

Sue: Looking at 11, 22, 33, 44, etc., I see that the difference between the units digit and the tens digit of every number is zero. Also, if I divide the tens by the units digit, I get one every time.

Mentor: Good observation. What do you think we might do to extend that thought out to 110, 121, and 132?

Sue: The number 110 resembles all the smaller numbers, say 88, and 99. The difference equals zero. The $1 - 1 = 0$ and the units zero does not change the formula. But the number 121 doesn't work the way I thought it should.

Mentor: Let me suggest numbering the digits, starting from the right with the 1 in 121 first position, 2 in the second position, and 1 in the third position. What result do you get, if you underline the digits in the odd positions, and subtract the sum of the even-positioned digits from the sum of the odd-positioned digits: 1<u>2</u>1?

Sue: $1 + 1 - 2 = 0$. This formula gives us the same zero result we had with all the two-digit numbers. Similarly, 11 divides 132, because $1 + 2 - 3 = 0$.

Mentor: We might note that 0 is a multiple of 11, just as 0 is a multiple of 9. Test 1727 for divisibility by 11. Abbie.

Abbie: 1<u>7</u>2<u>7</u> has the odd-positioned digits 7 and 7, adding up to 14. The even-positioned digits, 1 and 2, add up to 3. $14 - 3 = 11$. The difference is not zero, but eleven. Therefore, like the rule for 9, perhaps, eleven divides the number 1727.

Mentor: Perhaps, we will see why this test for the factor eleven works, if we learn the Trachtenberg method of multiplication by 11. We want to observe how multiplying by eleven follows a very simple pattern. Multiplying by 1 means just copy the number. Multiplying by 11 means copying the number twice, the second time shifted over one digit to the left. Let us multiply 7245 by 11. We copy the 7245 and add it to 72450, the 0 on the right shifting the number to the left. Adding the units digits, 5 and 0, copies the 5. Adding the tens digits, adds the neighbors 4 and 5. Continue, until we find ourselves adding the 7 on the left end to an unwritten 0, giving us 79695.

The traditional way of multiplying 11 times the number 7245 would put the two numbers on two different rows, lining up the units digits on the right.

|   | 7 | 2 | 4 | 5 |
|---|---|---|---|---|
|   |   |   | 1 | 1 |
|   |   |   |   |   |
|   |   |   |   |   |

In the first empty row under the 11, we enter the result of multiplying 7245 by 1, giving us 7245.

|   | 7 | 2 | 4 | 5 |
|---|---|---|---|---|
|   |   |   | 1 | 1 |
|   | 7 | 2 | 4 | 5 |
|   |   |   |   |   |

In the second empty row, we enter 10 times the number, 7245, or 72450, starting on the right side.

|   | 7 | 2 | 4 | 5 |
|---|---|---|---|---|
|   |   |   | 1 | 1 |
|   | 7 | 2 | 4 | 5 |
| 7 | 2 | 4 | 5 | 0 |
|   |   |   |   |   |

Next, we add each of the columns, starting on the right with 5 + 0, then 4 + 5, etc.

|   | 7 | 2 | 4 | 5 |
|---|---|---|---|---|
|   |   |   | 1 | 1 |
|   | 7 | 2 | 4 | 5 |
| 7 | 2 | 4 | 5 | 0 |
| 7 | 9 | 6 | 9 | 5 |

Notice, in each column under the eleven, we are adding neighboring digits in the number to be multiplied by 11. Let's try multiplying 8429 × 11, using the Trachtenberg method.

| 8 | 4 | 2 | 9 | × | 1 | 1 | = |   |   |   |   |

Sue: 8429 × 11 =     9, then adding the neighbors 2 and 9 for 11, recording the units 1 for 19, but I have an extra digit. What do I do with it?

| 8 | 4 | 2 | 9 | × | 1 | 1 | = |   |   | 1 | 9 |

Mentor: Anybody have a suggestion? George.
George: Carry the tens digit over to the next sum of neighbors, for 4 + 2 + 1 = 7. This would give us 719.

|   |   | 1 |   |   |   |   |   |   |   |   |
|---|---|---|---|---|---|---|---|---|---|---|
| 8 | 4 | 2 | 9 | × | 1 | 1 | = |   | 7 | 1 | 9 |

Mentor: Do you want to complete the multiplication with that suggestion in mind?
Sue: Now we have to add the 8 and the 4 to get 12, putting down the 2, for 2719, and carrying the 1. There is no digit to add the 1 to, so we just write it down. This gives us: 18429.

| 1 |   | 1 |   |   |   |   |   |   |   |   |
|---|---|---|---|---|---|---|---|---|---|---|
| 8 | 4 | 2 | 9 | × | 1 | 1 | = | 9 | 2 | 7 | 1 | 9 |

Mentor: What is our rule for multiplying by 11? Sue.
Sue: <u>Copy the units digit, adding the neighbors from right to left and carrying any tens digit to the next pair of neighbors. Finally, copy the left-most digit, adding any necessary carry.</u>
Mentor: Now I think you can all multiply by 11, and perhaps see why our rule for factors of eleven works. Notice that when we multiply by eleven, we add the neighboring digits and when we check for the factor eleven we are subtracting neighbors.

This completes our list of factor-identification rules for the time being.

**Between the Scenes**
Leaping ahead, traditional multiplication uses columns for units, tens, etc. digits. We show the multiplication of 427 × 64, placing each product's units digits below and its tens digits 'carry' above and to the left. 4 × 7 = 28, 8 below and 2 above to the left.

| Carry from 6 × |   | 2 | 1 | 4 |   |
|---|---|---|---|---|---|
| Carry from 4 × |   |   | 1 | 2 |   |
| First Factor |   |   | 4 | 2 | 7 |
| Second Factor |   |   |   | 6 | 4 |
| 4 times 427 = |   | 1 | 7 | 0 | 8 |
| 60 times 427 = | 2 | 5 | 6 | 2 | 0 |
| Product | 2 | 7 | 3 | 2 | 8 |

Using this system of tables with cells, multiply 851 × 97.

**Act 4**    **Operations on Numbers**

**Scene 1**    **Squares And Their Roots**

Mentor: To do and to understand math, we have to be able to multiply numbers, that is, find the products of numbers. From our table we can read on row 4 and column 5 the number 20. We say that 4 times 5 equals 20. Because you know how to count so many different ways, you can build your own multiplication table any time you want. This explains why our tables look the way they do, but we don't want to start all over again building a huge table just to calculate that 6 times 7, or 6 × 7, equals 42. So, again we look for special patterns, patterns helping us to remember the multiplication tables and giving us new insight at the same time.

We take pieces out of our new table to build another one, another T. We put the listed items on the left side of the T and their **squares** on the right. You will want to learn these unique values. I will say the number on the left and you the one on the right. Fill in the table as we go.

(The mentor gives every student a chance to recite a square.)
Mentor: 1?   Class: 1   Mentor: 2?   Class: 4   Mentor: 3?   Class: 9
Mentor: 4?   Class: 16   Mentor: 5?   Class: 25   Mentor: 6?   Class: 36
Mentor: 7?   Class: 49   Mentor: 8?   Class: 64   Mentor: 9?   Class: 81
Mentor: 10?   Class: 100   Mentor: 11?   Class: 121   Mentor: 12?   Class: 144

| number | number$^2$ |
|---|---|
| 1 | 1 |
| 2 | 4 |
| 3 | 9 |
| 4 | 16 |
| 5 | 25 |
| 6 | 36 |
| 7 | 49 |
| 8 | 64 |
| 9 | 81 |
| 10 | 100 |
| 11 | 121 |
| 12 | 144 |

Mentor: Notice we have a new symbol on the top of our T. We read 'number$^2$' as "number squared" where the 2 sits half a line higher than the word 'number.' We call the raised 2 an **exponent** and will talk about them later.

| number | number² |
|--------|---------|
| 1 | 1 |
| 2 | 4 |
| 3 | 9 |
| 4 | 16 |
| 5 | 25 |
| 6 | 36 |
| 7 | 49 |
| 8 | 64 |
| 9 | 81 |
| 10 | 100 |
| 11 | 121 |
| 12 | 144 |

Mentor: Now why do you suppose we call the numbers on the right side "the square" of the number on the left? We resort to visual images. What can we say about the adjacent sides of a **square**? They are equal and at right angles. See the square below.

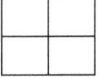

How many rows does this square have?
Sarah: 2
Mentor: How many columns does this square have?
Alexander: 2
Mentor: How many cells do we see?
Tom: 4, which equals 2 times 2!
Mentor: Excellent. Now, if I asked you how many cells are found in a square of 3 rows and 3 columns, what would you say?
Mary: 9, because 3 times 3 equals 9.
Mentor: So if I ask for the area of a square with 9 rows and 9 columns, how many cells would you find in the square?
Mary: $81 = 9 \times 9$

Mentor: We call 9 the **square root** of 81 ($\sqrt{81}$), or simply the **root** of 81. Now we can make a square root table. All we have to do is exchange the two columns. Sarah, read the right column of our T above so we know what to put in the left column.
Sarah: (Reading and entering in her T)

Mentor: Tom, tell us how to fill the right-hand column.
Tom: (Reading and entering, finishing the table below)

| number | √number |
|--------|---------|
| 1 | 1 |
| 4 | 2 |
| 9 | 3 |
| 16 | 4 |
| 25 | 5 |
| 36 | 6 |
| 49 | 7 |
| 64 | 8 |
| 81 | 9 |
| 100 | 10 |
| 121 | 11 |
| 144 | 12 |

**Between the Scenes**

1. Calculate the values of the squares of: 13, 14, and 15. What methods might you use to do the calculations?

2. What can you say about the square root of 12? If we had an approximation for it, where would it appear in our table? Where would we find the square root of 105?

3. We can look at $\sqrt{36}$ two different ways: 1) $\sqrt{6^2} = 6$ as we have introduced square roots, or 2) $\sqrt{4 \times 9} = \sqrt{4} \times \sqrt{9}$, observing that $\sqrt{4} \times \sqrt{9} = 2 \times 3$, or 6. A sentence describing this second observation asserts: **The (square) root of a product equals the product of the (square) roots**. How many other examples of this 'rule' can you find in our above list? (Do not consider the case that $1^2 = 1$.)

4. Show that 'the root of the product equals the product of the roots' each for 196 and 225.

**Act 4     Scene 2     Rectangles**

Mentor: In this scene, we calculate the areas of a few **rectangles**. As we have seen in Scene 1, a square's adjacent sides are equal and at right angles. How do we define a rectangle? Somebody want to guess?
Tom: It is a box, like a square, but the adjacent sides are perpendicular but not equal.
Mentor: Good. So, if we had a rectangle with two rows and three columns, how many cells are in the rectangle?
Tom: We could draw the rectangle and count the cells.
Mentor: OK! Everybody draw a rectangle with two rows and three columns. Let's enter the number of cells, starting at the upper left hand corner.

| 1 | 2 | 3 |
|---|---|---|
| 4 | 5 | 6 |

Mentor: How could we calculate the number of cells faster than drawing a rectangle and counting all the cells?
Tom: Just multiply 2 times 3.
Mentor: Excellent. We multiply the number of the rows times the number of columns. What would we do with the rods to get these answers?
Sarah: Make two rows of three white rods, and count them. I also find that I can make two rows of green rods or three columns of red rods and come out with equal areas!

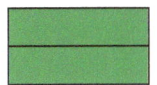 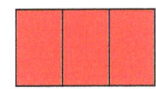

Mentor: Here we have the observation that $2 \times 3 = 3 \times 2$. We call this the **symmetric** property when the numbers can change sides of an operation like multiplication. So we see we have many ways of multiplying two numbers: counting the number of rods in a rectangle, counting the number of cells, counting by one of the numbers the other number of times, or learning the multiplication tables "by heart." Which method will be fastest?
Alexander: By heart!
Mentor: We should all try to learn the multiplication tables by heart to speed up our work, but we should be able to recreate any result we might have forgotten by one of the methods above.

**Between the Scenes**
1.     Display a rectangle 5 white rods by 8 white rods. How many rods in the rectangle?
2.     How many white rods are in a rectangle 3 by 11?
3.     Can you name another symmetric operation? Several **asymmetric** (non-symmetric) operations?

**Act 4     Scene 3     Divide And Conquer**

Mentor: We have mentioned four operations so far: addition, subtraction, multiplication, squaring. This time we will look at **division**. First, let me ask you what we mean when we say, "Let's divide the birthday cake so everybody gets a piece."
Tom: We will cut the cake up into as many pieces as there are people.
Mentor: Do you assume anything about the size of the pieces?
Tom: Yes. Mom always wants a small piece and Dad a large piece, with everybody else getting roughly equal pieces!
Mentor: So we now have a new operation on numbers: **division**. One number divides another, if I can make one-color train of the first number the same length as the second number's train.

For instance, red (2) divides dark green (6), because three red rods makes a train equal to a dark green train. We say 2 divides 6 three times, or 6 divided by 2 equals 3, or 2 times 3 equals 6. Notice that we express division, using the rods, the same way we express multiplication!

Can anybody name another pair of numbers, the one dividing the other?
Jane: Purple divides brown into two pieces.
Mentor: What are you saying, Jane, about numbers?
Jane: 4 divides 8.
Mentor: What do we get when we divide 8 by 4?
Jane: 2.
Mentor: Can anyone tell us a number that divides 12?
Jane: 2, 3, 4 or 6.
Mentor: Good. Notice we represent the number 12 with two rods: one orange and one red. So the trains do not have to be a one-color train. We say the first train divides the second if we can make a train of one or more first trains the same length as the second. Does 12 divide 24?
Mary: Yes. I can make a train of two orange-and-red trains equal to a train of two orange rods and one purple one, so 12 divides 24 two times.
Mentor: Can I always divide one number into another?
George: No. 3 does not divide 5!
Mentor: Excellent. We know we can divide an all blue train by 9 by counting the rods. If we add two blue trains, can we divide the resulting sum train by 9?
Jane: Yes, the sum will be a blue train, so we just count the number of blue rods.
Mentor: Good. We have made a lot of progress!

**Between the Scenes**
1.     What numbers divide 15? 24? 29? How can you show the correctness of your answers?
2.     What do you have to do to find all the factors of these numbers?
3.     If you add a blue train and a brown train, must the sum train be divisible by 9 and by 4? By 9 or by 4? If a brown and blue train is divisible by 9, how many brown rods must it contain?

**Act 4**      **Scene 4**      **Lattice Multiplication**

Mentor: When we built our multiplication table, we limited our numbers to one digit. Suppose we had to multiply 258 times 73. How could we do that? We have a method called **Lattice Multiplication**, using an array of cells as shown below. Copy my drawing on your own piece of paper!
    We make our rectangle two cells wider than the number of digits in our first number: one on the left and one on the right. Here we have five columns. In a similar fashion, we make four rows by placing one row above the 73 and one below. Next we draw diagonals from the upper right to the lower left of the cells below the digits of 2, 5, and 8 and to the left of the digits 7 and 3.

Our first lattice.

Now we will multiply each digit in the one number by the digits of the other number, placing the product number in the cell in the column and row as we did in our multiplication tables. So in the column under the 2 and the row containing the 7 we write the product, 14. We use the diagonal to separate the two digits of the product: the tens digit above the diagonal and the units digit below. Here we have the 1 in the upper left corner of the product cell and the 4 in the lower right corner.
    Sarah, please tell us how to fill in the next cell under the five and to the left of the 7.
Sarah: 5 times 7 equals 35. I will put the 3 above the diagonal and the 5 below.
Mentor: Excellent. George, the next cell.
George: 8 times 7 equals 56. The 5 goes above, the 6 below the diagonal.
Mentor: Jane, how will we fill the cell under the two and to the left of the 3?
Jane: I only have one digit, a 6.
Mentor: Is the 6 a units digit or a tens digit?
Jane: Units. Oh, so place it under the diagonal.

Mentor: Yes. How can we indicate a missing tens digit?
Jane: With a zero.
Mentor: Exactly!

(The Mentor continues around the class until all the triangles contain the correct numbers.)

Mentor: We start building the product from the lowest right product cell, outlined in red with the 2 above and the 4 below the diagonal. We simply record the units digit, the underlined 4, in the cell directly below, as shown.

Mentor: Our next step consists of adding the three orange digits between the two diagonals on the lower right side: 6 + 2 + 5 = 13. We have a two-digit number but only record the units digit, the 3, in the cell to the left of the 4. So what should we do with the 1? We carry it over to the next sequence of digits (5, 5, 1, 6) between two diagonals.

We add these four green digits and our carried number, 1. So 5 + 5 + 1 + 6 + 1 = 18, with the last 1 being the carry. Put the 8 below the 6.

|   | 2 | 5 | 8 |   |
|---|---|---|---|---|
|   | 1 / 4 | 3 / 5 | 5+1 / 6 | 7 |
|   | 0 / 6 | 1 / 5 | 2 / 4 | 3 |
|   | 8 | 3 | 4 |   |

Mentor: Alexander, what do you think we will do with the 1 from the 18?
Alexander: Carry it to the red sequence between the next pair of diagonals.
Mentor: Perfect. What do we add up and where do we put the result?
Alexander: We add 1 + 3 + 4 + 0 = 8 and put it next to the other 8.
Mentor: You added the numbers correctly but put the result in the wrong box. We use the box, in red, at the bottom of the pair of diagonals to record our result.

|   | 2 | 5 | 8 |   |
|---|---|---|---|---|
|   | 1 / 4 | 3+1 / 5 | 5+1 / 6 | 7 |
| 8 | 0 / 6 | 1 / 5 | 2 / 4 | 3 |
|   | 8 | 3 | 4 |   |

Mentor: Finally, Mary, where do we go next?
Mary: There being no carry, we take the purple 1 from above the last diagonal and record it off to the left.

|   | 2 | 5 | 8 |   |
|---|---|---|---|---|
| **1** | 1/4 | 3/5 | 5/6 | 7 |
| 8 | 0/6 | 1/5 | 2/4 | 3 |
|   | 8 | 3 | 4 |   |

Mentor: And we have completed the multiplication, but how do we read it? Tom!
Tom: We read first down from the top left and then across the bottom, getting 18,834.
Mentor: How did you know to do that?
Tom: I don't know.

Mentor: Class, did you learn something new?   Yes. What?
Mentor: Class, did you have some fun?   Yes. Why?

**Between the Scenes**
1. Multiply by lattice multiplication 348 times 729. Talk us through the process.
2. Use lattice multiplication to calculate the product of 48 times 7,319.

**Act 4**         **Scene 5**              **Conditional Replacement**

Mentor: If I have a one-color purple train, can I make another one-color train equal to it but with a different color?
Mary: Yes, white.
Mentor: Good. So you could replace my purple train with a white train, no matter how long my purple train?
Mary: Yes. Always!
Mentor: Mary, can you think of another one-color train that always equals my purple train?
Mary: Yes, a red train.
Mentor: So any purple I have you can replace with an equal red train? How do you understand that?
Mary: I will replace each purple rod with two red ones. When all the purple rods disappear, I will have replaced the purple train with a red train.
Mentor: Well thought out! Now think about the reverse. Can you replace *every* red train with a purple train?
Mary: I think so. If I have a red train as long as a brown train, then I can replace pairs of red rods with purple rods.
Mentor: That example supports your statement, but tell me about a red train that equals a dark green rod. Can you replace this red train with purple rods?
Mary: Oops! No. So there exists one red train I cannot replace with a purple train.
Mentor: Try replacing an orange train with a red and a purple train.
Mary: I cannot do that. So there exist two red trains I cannot replace with purple trains. Hmmm. I see. I cannot replace a train of one orange and one purple rod with purple rods, but I can with red rods! So we have many red-rod trains we cannot replace with purple-rod trains.
Mentor: Yes. We may replace every purple train by a red train, but not every red train by a purple train. Can anybody think of another color besides purple that we can always replace with red trains but cannot always replace a red train?
Alexander: I can always make a red train the same length as a dark green train, but I cannot make a dark green train the same length as a train of two red rods.
Mentor: Excellent. In terms of numbers, we say 2 (red) divides 6 (dark green), but 6 does not divide 2. Does 2 divide 10?
Alexander: Yes, I can always make a red train equal to an orange train, but I cannot divide 2 by 10, because I cannot replace a single red rod by an orange one.

Mentor: We can now say <u>if a dark green train divides another train, then a red train divides the second train.</u> But we cannot say if a red train divides another train, then a dark green train divides the other train.
    In terms of numbers, if six divides a number then 2 divides the number. But we cannot say if two divides a number, then 6 divides that number. Question: If eight divides a certain number, what other numbers divide the certain number? 8 divides 56, so what other numbers must divide 56, besides 7?

Sarah: If 8 divides 56, then 2 and 4 must divide 56, because they divide 8.
Mentor: Good thinking! Our next scene fragment takes a more verbal approach.

(This final part of the Scene may suit older students better than younger.)
Mentor: This part of our study of mathematics extends **conditional reasoning**. An example of this kind of thinking states: I promise you that if you are well behaved, then you will get a cookie. There are two conditions: (1) you are well behaved and (2) you are not well behaved. There are also two conclusions: (1) you get a cookie and (2) you do not get a cookie.

We want to know when you will think I did not keep my promise: (1) You behave well and you get a cookie or (2) you behave well and you do not get a cookie?
Tom: (1) If I behave I should get my cookie! (2) I will feel disappointed that you did not keep your promise in the second case: good behavior but no cookie.
Mentor: OK. Here comes the weird part. Now the second condition: (1) You misbehave and do get your cookie or (2) you misbehave and do not get a cookie?
Tom: I did not expect to get the cookie in this situation (1), so I am pleased, even surprised to get a cookie, not disappointed!
Mentor: And the last case, where you misbehave and don't get your cookie?
Tom: I would still like to get my cookie, but I am to blame that I did not get it. No reasonable disappointment.
Mentor: You have the idea. We have four different situations to consider in conditional sentences: (1) the condition (the "if" part) is true and the conclusion (the "then" part) is true, (2) the condition is true and the conclusion is false, (3) the condition is false and the conclusion is true, (4) the condition is false and the conclusion is false.

| If\Then | Cookies | No Cookies |
|---|---|---|
| Good | (1) | (2) |
| Bad | (3) | (4) |

Did you learn something new? Have some fun?

**Between the Scenes**
1. Can you always replace an orange train by a yellow one? Can you always replace a yellow train by an orange one?
2. Consider the promise: If you get an 'A' in math, then I will buy you new shoes. What are the four situations? When will you have reason to feel cheated? When will you be surprised to get the new shoes?
3. Create your own conditional (if-then) sentence and discuss the possible situations. When will your consider the sentence false?

**Act 4      Scene 6      Reaching The Remainders**

Mentor: We have dealt with multiplication in such a manner that you can multiply any two **natural numbers**, those numbers we use for counting, starting at 1 and 2 and on up. We have discovered that when we multiply any two numbers, we can divide the product by either of the two original numbers. But can we divide any natural number by any other natural number? You know we cannot divide 7 by 2. 7 is odd and 2 is even. We cannot divide 2 by 7, because 7 is bigger than 2, and any black train will be longer than a single red rod!

Let's consider a division problem we have already solved. Does 5 divide 15?
Tom: Five divides 15, because a train of three yellow rods equals a train of one orange plus one yellow rod.

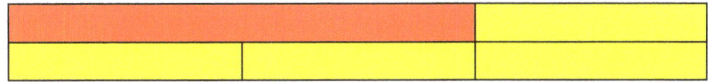

Mentor: Does 3 divide 15 also? How would you show me it does? Mary.
Mary: We could count the three yellow rods in the train above, or we could construct a green train of the same length. See?

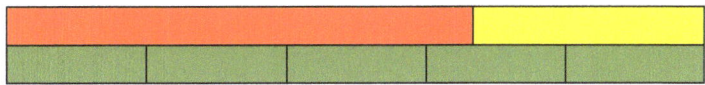

Mentor: What equation expresses this situation?
Mary: O + Y = G + G + G + G + G.
Mentor: Tired of writing G's? (Mary nods.) How many green rods do we have?
Mary: 5.
Mentor: So, if I wrote O + Y = 5G, how would you understand that?
Mary: Orange plus yellow equals five greens.
Mentor: Good. So, we indicate multiplying 5 times 3 using green rods with "5G."

Does 3 divide 16 also?
Mary: No. An orange and a dark green train would require us to add a white rod to our all-green train. Like this!

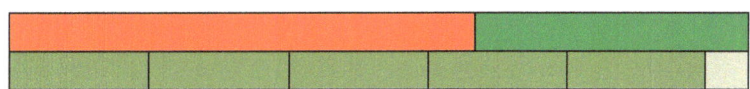

Mentor: What equation expresses this new arrangement?
Mary: O + D = 5G + W, or 16 = 5 × 3 + 1.

Mentor: Are you thinking of multiplying 5 times 3 first and then adding 1, or adding 3 and 1, then multiplying by 5?
Mary: Multiplying first, reading from left to right.
Mentor: What would you do with $1 + 5 \times 3$?
Mary: Hmm. If I add 1 and 5, to get 6, and multiply by three, I will get 18, not the 16 we want.
Mentor: You see the problem correctly. Our understanding of a mixture of addition and multiplication states: <u>multiplication always comes before the addition</u>. In this way we get the 16 we want, starting from the left or the right.

We call the white rod you added to make a train equal to sixteen, the **remainder** when three divides sixteen. We also say that three does not go into sixteen an even number of times.
What remainder do we get, if we divide seventeen by 3? Alexander.
Alexander: I would have to put a red rod where the white one was. So the remainder would be 2. Here is my proof!

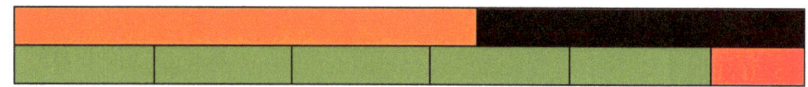

Mentor: Can you give us an equation for this pair of trains?
Alexander: $O + K = 5G + R$, or $17 = 5 \times 3 + 2$.

Mentor: Exactly. What happens with 18, Jane?
Jane: 3 divides 18, so we do not have a remainder.
Mentor: Yes. Can you give us an equation?
Jane: $O + N = 6G$, or, with numbers: $18 = 6 \times 3$.
Mentor: What number could we add to the right side of your equation and have no change?
Jane: How about a zero?
Mentor: What would the equation look like?
Jane: $18 = 6 \times 3 + 0$.

Mentor: Does 3 divide 19? Joe.
Joe: No. The green train would require an additional white rod, just like sixteen did. See my parallel trains!

Mentor: So what does your equation look like?
Joe: $19 = 6 \times 3 + 1$.
Mentor: Exactly! Stepping up one higher, what happens when we divide 3 into 20?
George: We get $20 = 6 \times 3 + 2$.
Mentor: And with 21?

George: We know 7 × 3 = 21, so 21 = 7 × 3 + 0.
Mentor: What pattern do you see here?
George: Dividing by 3, we get remainders of 0, 1, and 2, repeating again and again.
Mentor: Excellent observation. What remainders would you expect dividing natural numbers by 4? Why?
George: I think we will find four remainders. If we divide 8 by 4, we get a remainder of zero, 9 by 4 requires a white filler to give us a remainder of 1, 10 by 4 produces 2, 11 by 4 a 3, and 12 by 4 takes us back to zero. We get the remainders 0, 1, 2, 3 when we divide by 4.
Mentor: Nice explanation. Anybody want to tell us the remainders when we divide by 5? Sarah.
Sarah: 0, 1, 2, 3, 4.
Mentor: Why?
Sarah: For all those numbers we used counting by fives, we get a zero remainder, and each number in between 5's the remainders go up by one from one to four.
Mentor: In the table below, we see 5 divided into each of the numbers from 5 to 10, with the numerical remainders indicated in the right hand column.

| 1 | 2 | 3 | 4 | 5 | 6 | 7 | 8 | 9 | 10 | Remainders |
|---|---|---|---|---|---|---|---|---|----|------------|
|   |   |   |   |   |   |   |   |   |    |            |
|   |   |   |   |   |   |   |   |   |    | 0          |
|   |   |   |   |   |   |   |   |   |    |            |
|   |   |   |   |   |   |   |   |   |    | 1          |
|   |   |   |   |   |   |   |   |   |    |            |
|   |   |   |   |   |   |   |   |   |    | 2          |
|   |   |   |   |   |   |   |   |   |    |            |
|   |   |   |   |   |   |   |   |   |    | 3          |
|   |   |   |   |   |   |   |   |   |    |            |
|   |   |   |   |   |   |   |   |   |    | 4          |
|   |   |   |   |   |   |   |   |   |    |            |
|   |   |   |   |   |   |   |   |   |    | 0          |

Mentor: Did you learn something new?         Yes. What?
Mentor: Did you have some fun discovering it?   Yes. Why?

**Between the Scenes**
1. What are the possible remainders when we divide a number by 11?
2. Where do we use remainders in everyday life?

**Act 4        Scene 7            Clocks**

Mentor: A (trick) question: How much is the sum of 11 + 5? (Pause)
Alexander: 16.
Mentor: No, it's 4.
Alexander: Uh?
Mentor: If the time is 11, what time will it be in 5 hours?
Alexander: 4 o'clock.
Mentor: How do we arrive at that number?
Alexander: We add 5 to 11 and then subtract 12.
Mentor: So, what is the time six hours after 10 o'clock? Sarah.
Sarah: 4 o'clock.

Mentor: What time will it be in 12 hours, if the clock says 7 o'clock now?
Tom: The clock will indicate 7 o'clock again.
Mentor: So, adding twelve is the same as adding what other number?
Tom: Zero.
Mentor: Exactly. Let's make out an addition table for clock arithmetic.
(Handing out graph paper and instructing the students how to put the lists across the top and down the left side with a plus sign in the upper left hand corner. The Mentor gets a different student to fill in each row.)

| +  | 1  | 2  | 3  | 4  | 5  | 6  | 7  | 8  | 9  | 10 | 11 | 12 |
|----|----|----|----|----|----|----|----|----|----|----|----|----|
| 1  | 2  | 3  | 4  | 5  | 6  | 7  | 8  | 9  | 10 | 11 | 12 | 1  |
| 2  | 3  | 4  | 5  | 6  | 7  | 8  | 9  | 10 | 11 | 12 | 1  | 2  |
| 3  | 4  | 5  | 6  | 7  | 8  | 9  | 10 | 11 | 12 | 1  | 2  | 3  |
| 4  | 5  | 6  | 7  | 8  | 9  | 10 | 11 | 12 | 1  | 2  | 3  | 4  |
| 5  | 6  | 7  | 8  | 9  | 10 | 11 | 12 | 1  | 2  | 3  | 4  | 5  |
| 6  | 7  | 8  | 9  | 10 | 11 | 12 | 1  | 2  | 3  | 4  | 5  | 6  |
| 7  | 8  | 9  | 10 | 11 | 12 | 1  | 2  | 3  | 4  | 5  | 6  | 7  |
| 8  | 9  | 10 | 11 | 12 | 1  | 2  | 3  | 4  | 5  | 6  | 7  | 8  |
| 9  | 10 | 11 | 12 | 1  | 2  | 3  | 4  | 5  | 6  | 7  | 8  | 9  |
| 10 | 11 | 12 | 1  | 2  | 3  | 4  | 5  | 6  | 7  | 8  | 9  | 10 |
| 11 | 12 | 1  | 2  | 3  | 4  | 5  | 6  | 7  | 8  | 9  | 10 | 11 |
| 12 | 1  | 2  | 3  | 4  | 5  | 6  | 7  | 8  | 9  | 10 | 11 | 12 |

Mentor: Notice that we could have replaced every 12 by 0. So our table would look more like an ordinary addition table, but we can find some new patterns.
    Could someone tell us of a new pattern? Sarah.
Sarah: Adding 7 and 8 gives us 3, the same as adding 8 and 7.
Mentor: Do you have another pair of numbers behaving this way?

Sarah: Yes. It works for all the numbers, adding the first to the second number equals the same as adding the second to the first.
Mentor: Do you know a name for this property?
Sarah: No.
Mentor: Suppose we draw a line from the upper left corner to the lower right, what do you see on the two sides of the line?
Sarah: The same thing.
Mentor: Have you done something today, before coming to school, and had a similar experience?
Sarah: Hmmm. Oh, I looked in the mirror to comb my hair.
Mentor: When you see on the right the same thing that is on the left, we call this the **symmetric** property.

Now we will explore the multiplication of clock numbers. What do we get when we multiply 4 times 5 in clock arithmetic? George.
George: We know 4 times 5 equals 20, and 20 less 12 equals 8.
Mentor: Can you find this result another way? How did we build our first, regular multiplication table?
George: We counted.
Mentor: So take us through the process.
George: 5 + 5 = 10, and 10 + 5 = 15, and 15 – 12 = 3, and 3 + 5 = 8.
Mentor: Fine. Let's fill out a multiplication table for clock arithmetic. We could also call 'clock' arithmetic **'remainder'** or **'modular' arithmetic**.

| × | 1 | 2 | 3 | 4 | 5 | 6 | 7 | 8 | 9 | 10 | 11 | 0 |
|---|---|---|---|---|---|---|---|---|---|----|----|---|
| 1 | 1 | 2 | 3 | 4 | 5 | 6 | 7 | 8 | 9 | 10 | 11 | 0 |
| 2 | 2 | 4 | 6 | 8 | 10 | 0 | 2 | 4 | 6 | 8 | 10 | 0 |
| 3 | 3 | 6 | 9 | 0 | 3 | 6 | 9 | 0 | 3 | 6 | 9 | 0 |
| 4 | 4 | 8 | 0 | 4 | 8 | 0 | 4 | 8 | 0 | 4 | 8 | 0 |
| 5 | 5 | 10 | 3 | 8 | 1 | 6 | 11 | 4 | 9 | 2 | 7 | 0 |
| 6 | 6 | 0 | 6 | 0 | 6 | 0 | 6 | 0 | 6 | 0 | 6 | 0 |
| 7 | 7 | 2 | 9 | 4 | 11 | 6 | 1 | 8 | 3 | 10 | 5 | 0 |
| 8 | 8 | 4 | 0 | 8 | 4 | 0 | 8 | 4 | 0 | 8 | 4 | 0 |
| 9 | 9 | 6 | 3 | 0 | 9 | 6 | 3 | 0 | 9 | 6 | 3 | 0 |
| 10 | 10 | 8 | 6 | 4 | 2 | 0 | 10 | 8 | 6 | 4 | 2 | 0 |
| 11 | 11 | 10 | 9 | 8 | 7 | 6 | 5 | 4 | 3 | 2 | 1 | 0 |
| 0 | 0 | 0 | 0 | 0 | 0 | 0 | 0 | 0 | 0 | 0 | 0 | 0 |

Mentor: Let's fill in row 2 together. Tom.
Tom: 2, 4, 6, 8, 10, 0, 2, 4, 6, 8, 10, 0.
Mentor: Have you seen this sequence before?
Tom: Of course, we are counting by 2's up to 12 and then starting all over again.
Mentor: What do you find weird about this table?
Tom: I can multiply two numbers, 2 times 6, and get zero.
Mentor: Why do you find this weird?
Tom: If I multiply 6 times zero, I get zero. Or if I multiply 2 times zero, I expect to get zero, because we are just adding six zeros or two zeros. But neither two and nor six is zero yet we get a zero product.
Mentor: Nice observation. We are working here with clock arithmetic. What kind of numbers do we have?
Tom: They are remainders after we divide by 12. We get zero with other numbers: 3 times 4, and 8 times 3, and 9 times 4.
Mentor: Exactly! What are the squares here? Sarah.
Sarah: 1, 4, 9, 4, 1, 0. I see symmetry here.
Mentor: What do you mean?
Sarah: 1, 4, 9, 4, 1 is symmetric about 9, and 1, 4, 9, 4, 1, 0, 1, 4, 9, 4, 1 is symmetric about 0.
Mentor: Beautiful! Let's move on to the next scene to look at a clock with just seven hours. We want to see what kind of patterns we can find in this new "clock."

Mentor: Something new?   Yes. What?
　　　　Some fun?   　　　Yes. Why?

**Between the Scenes**

1.　If our twelve-hour clock indicates 5 o'clock and we have to wait two eight-hour periods until our next opportunity to see our friend, at what time will we see her?

2.　In the military, it is customary to use a 24-hour clock. What benefits does such a time system serve? In Europe train schedules are usually given in 24-hour units.

3.　Why do you suppose 12 and 24 were used originally to keep track of the hours of the day?

**Act 4**  **Scene 8**  **Modular Arithmetic**

Mentor: In the last scene, we took our arithmetic from the clock on the classroom wall. That clock repeats itself every 12 hours, but our new "clock" will repeat itself every seven "hours." Perhaps we should imagine ourselves on another planet or moon that rotates faster than our earth. Out on Moon X the local people recognize that 7 divided by 7 gives a remainder of 0, so they use 0 instead of 7. Let's take out our graph paper and write our two lists (0, 1, 2, 3, 4, 5, 6), one horizontal, the other vertical, with a "+" sign in the upper left hand cell.

| + | 0 | 1 | 2 | 3 | 4 | 5 | 6 |
|---|---|---|---|---|---|---|---|
| 0 |   |   |   |   |   |   |   |
| 1 |   |   |   |   |   |   |   |
| 2 |   |   |   |   |   |   |   |
| 3 |   |   |   |   |   |   |   |
| 4 |   |   |   |   |   |   |   |
| 5 |   |   |   |   |   |   |   |
| 6 |   |   |   |   |   |   |   |

Mary, would you please tell us what to write in the first row, the zero row.
Mary: 0, 1, 2, 3, 4, 5, 6.
Mentor: OK. Try the next row as well.
Mary: 1, 2, 3, 4, 5, 6, 0.
Mentor: Great. Next row, George.
George: 2, 3, 4, 5, 6, 0, 1.
Mentor: That's the idea. Now everybody fill in the rest of the addition table.

| + | 0 | 1 | 2 | 3 | 4 | 5 | 6 |
|---|---|---|---|---|---|---|---|
| 0 | 0 | 1 | 2 | 3 | 4 | 5 | 6 |
| 1 | 1 | 2 | 3 | 4 | 5 | 6 | 0 |
| 2 | 2 | 3 | 4 | 5 | 6 | 0 | 1 |
| 3 | 3 | 4 | 5 | 6 | 0 | 1 | 2 |
| 4 | 4 | 5 | 6 | 0 | 1 | 2 | 3 |
| 5 | 5 | 6 | 0 | 1 | 2 | 3 | 4 |
| 6 | 6 | 0 | 1 | 2 | 3 | 4 | 5 |

Mentor: Next we have the multiplication table: the same setup, just an "×" (multiplication sign) in the upper left hand cell. What do we write in the first row,
Tom: 0 seven times.
Mentor: Correct. Since that was so easy, try the next row as well.
Tom: 0, 1, 2, 3, 4, 5, 6.
Mentor: Tough, eh? The two's row, Jane.

Jane: 0, 2, 4, 6, 1, 3, 5.
Mentor: OK. The 3's row, Sarah.
Sarah: 0, 3, 6, 2, 5, 1, 4.
Mentor: How can we speed up the process?
Sarah: The three's column will be the same as the three's row. So we can copy the last numbers in the row in the column.
Mentor: Good. George, the 4's.
George: Fill in 2, 6, 3 in columns 4, 5, 6, respectively. Also, 6 and 3 in the 4's column.

| × | 0 | 1 | 2 | 3 | 4 | 5 | 6 |
|---|---|---|---|---|---|---|---|
| 0 | 0 | 0 | 0 | 0 | 0 | 0 | 0 |
| 1 | 0 | 1 | 2 | 3 | 4 | 5 | 6 |
| 2 | 0 | 2 | 4 | 6 | 1 | 3 | 5 |
| 3 | 0 | 3 | 6 | 2 | 5 | 1 | 4 |
| 4 | 0 | 4 | 1 | 5 | 2 | 6 | 3 |
| 5 | 0 | 5 | 3 | 1 | 6 | 4 | 2 |
| 6 | 0 | 6 | 5 | 4 | 3 | 2 | 1 |

Mentor: What property do we employ to fill that column, George?
George: Symmetry.
Mentor: Excellent. Complete the bottom right-hand four cells, Alexander.
Alexander: $5 \times 5 = 25$, and $25 - 21 = 4$, and so $5 \times 6 = 30$, and $30 - 28 = 2$.
   Repeat those calculations for column 5. $6 \times 6 = 36$, and $36 - 35 = 1$.
Mentor: What strikes you right away as different from the 12-hour table?
Alexander: There are no zeros in the middle of the new table.
Mentor: Can you guess the reason why?
Alexander: 12 is a composite number, but 7 is prime. So seven does not have any factors other than 1 and 7.
Mentor: Good observation. Can anyone tell us where we use modular arithmetic everyday?
Tom: My dad told me the computers use this kind of arithmetic with just 0's and 1's.
Mentor: Yes. Why?
Tom: I don't know why.
Mentor: Electronic computers consist of millions of switches. These switches are either on or off. We might think of on as "1" and off as "0." Then we have a very small table with just two numbers. Now that we have had a brief view of some numbers we don't usually think about, we will go back to numbers we use frequently but perhaps do not understand well: fractions.

**Between the Scenes**
1. Looking at our seven-hour clock, we see 3 o'clock. What time will we see on the clock in 10 hours?
2. Our seven-hour clock indicates 5 o'clock. What time will we see in 4 hours?

**Act 5**  **Fractions and Exponents**

**Scene 1**  **Fractional Parts**

Mentor: "I want my half." and "Each of you gets a third." are claims or statements heard at home and elsewhere. We want to look at these expressions a little more closely than we usually do. You all can tell me the value of "half of four," but can you tell me how to express that idea mathematically? Oh, Alexander.

Alexander: "Half" looks like this: $\frac{1}{2}$.

Mentor: We can write it that way with a horizontal line: $\frac{1}{2}$. We want to write half of four, and this we express as a multiplication: $\frac{1}{2} \times 4$. We say: "one half times four."
What is half of four?
Alexander: 2.
Mentor: How can we express this idea with the rods?
Tom: We could make a red train parallel to a purple one. Each red rod would be half of the purple rod.

Mentor: Exactly. What is the value of two halves of four or $\frac{2}{2} \times 4$?

Sarah: 4.

Mentor: $\frac{3}{2} \times 4$?   Sarah: 6.   Mentor: $\frac{4}{2} \times 4$?   Sarah: 8.   Mentor: $\frac{5}{2} \times 4$?   Sarah: 10.

Mentor: We seem to have the idea of one half in mind. What about one third? Evaluate 1/3 of 3. Tom.
Tom: 1.
Mentor: Explain that.
Tom: A train of three white rods equals a parallel green train.
Mentor: Evaluate two thirds of 3.   Tom: 2.   Mentor: 3/3 of 3.   Tom: 3.
   Mentor: 4/3 of 3.   Tom: 4.
Mentor: Let's calculate one third of 6. Sarah.
Sarah: For one third of 6, I need three red rods to make a train equal to a dark green train. One of those red rods, or 2, represents $\frac{1}{3} \times 6$.

Mentor: I'll give you an expression and you evaluate it. $\frac{2}{3} \times 6$?   Sarah: 4.

Mentor: $\frac{3}{3} \times 6$?  Sarah: 6   Mentor: $\frac{4}{3} \times 6$?   Sarah: 8   Mentor: $\frac{5}{3} \times 6$?   Sarah: 10.

Mentor: Let's try a quarter of something. $\frac{1}{4} \times 8$?  George: 2.   Mentor: $\frac{2}{4} \times 8$?

George: 4.    Mentor: $\frac{3}{4} \times 8$? George: 6.   Mentor: $\frac{5}{4} \times 8$? George: 10.

Mentor: We will try another number: 15. Calculate one fifth of fifteen, or $\frac{1}{5} \times 15$.

Alexander: $\frac{1}{5} \times 15 = 3$.    Mentor: So, $\frac{2}{5} \times 15 =$?    Alexander: 6

Mentor: $\frac{3}{5} \times 15 =$? Alexander: 9   Mentor: $\frac{4}{5} \times 15 =$? Alexander: 12.

Mentor: $\frac{6}{5} \times 15 =$? Alexander: 18   Mentor: $\frac{7}{5} \times 15 =$? Alexander: 21.

Mentor: Can we divide every brown train by 4?
Jane: Yes, we can always divide every brown train by 4, because we can replace each brown rod of the train by two purple rods, a multiple of 4.
Mentor: If we add one yellow rod to the brown train can we still divide the train by 4?
Jane: No. If we add on a yellow rod, we cannot divide the new train by 4, because we have the sum train, one part divisible by 4 and the rest not divisible by 4.
Mentor: Can we divide 29 by 4?
Jane: No.  A train of three brown rods equals 24 and adding one yellow rod makes it 29.  And we cannot divide 5 by 4.
Mentor: You have the idea.  Did you have some fun?  Did you learn something new?

**Between the Scenes**
(For the youngest students, it might be too confusing to introduce expressions like $\frac{2}{3} \times 5 = \frac{10}{3}$ this early in their arithmetic experience.  It will appear later.)

1. Examine the sequence of $\frac{1}{7} \times ( \ )$, where the parentheses contain 7, 14, 21, or 28, all multiples of 7.

2. Examine the sequence of $\frac{( \ )}{7} \times 14$, where the parentheses contain 1, 2, ..., 12, working in **improper fractions**, those with numerators larger than denominators.

3. Examine the sequence of $\frac{( \ )}{5} \times [ \ ]$, where the ( ) contain counting numbers starting at 1 up to 10, and the [ ] contain multiples of 5 such as 10, 15, 20, etc., always making sure that 5 divides the number in the [ ].

4. In the Between the Scenes following Act 4 Scene 1, we observed that 'the root of a product equals the product of the roots.'  Similarly, **the root of a quotient equals the quotient of the roots** (e.g. of the numerator and denominator).  Using this new observation, simplify $\sqrt{\frac{4}{9}}$ two ways.   $\sqrt{\frac{4}{9}} = \frac{2}{3}$ and $\sqrt{\frac{4}{9}} = \frac{\sqrt{4}}{\sqrt{9}}$ or $\frac{2}{3}$.  Repeat the process with $\sqrt{\frac{25}{16}}$, and $\sqrt{\frac{169}{196}}$, and $\sqrt{\frac{49}{225}}$.

**Act 5        Scene 2             Mixed Operations**

Mentor: We have calculated several expressions, combining addition and multiplication in Act 4 Scene 7. In this scene we will make the expressions a little more complicated, having two multiplications and one addition or subtraction. What did we learn earlier about an expression such as $3 \times 5 + 2$? Do we multiply the 3 and the 5 first and then add the 2, or do we add the 2 and the 5 before multiplying by 3? Mary.
Mary: We agreed to multiply first, even if the 2 came first as in $2 + 5 \times 3$. So we get 17, because $3 \times 5 = 15$ and $15 + 2 = 17$.

Mentor: Just so! Now what should we do with the expression: $3 \times 5 + 4 \times 2$? Alexander.
Alexander: Mary said we should multiply first, so $3 \times 5 = 15$ and $4 \times 2 = 8$. Then we add 15 and 8. $15 + 8 = 23$.
Mentor: You have the correct answer by following the rule: <u>multiplication comes before addition.</u> What would you do if the addition were made into a subtraction?
Jane: What should I calculate?
Mentor: $3 \times 5 - 4 \times 2$.
Jane: OK. We already know that $3 \times 5 = 15$ and $4 \times 2 = 8$, so all we have to do is subtract eight from fifteen: $15 - 8 = 7$.

Mentor: Now we change the numbers, asking if we have to change the process. George.
George: No. We will still do the <u>multiplication before the addition or subtraction</u>.
Mentor: Looking at $11 \times 6 - 4 \times 3$, what do we do? Mary.
Mary: $11 \times 6 = 66$ and $4 \times 3 = 12$, so we have to subtract 12 from 66.
Mentor: Continuing?
Mary: Yes. Subtract 10 from 66 for 56, and then subtract 2 more for 54!
Mentor: Mary, you have done it all! We can now return to our fractions in the next scene.
(The Mentor gives each child a sum or a difference of two products to be sure everybody grasps the idea.)

**Between the Scenes**
1.      Simplify each of the following:
 a) $7 \times 8 - 6 \times 5$      b) $5 \times 9 + 4 \times 11$      c) $7 \times 6 - 3 \times 8$      d) $6 \times 12 - 9 \times 7$
 e) $9 \times 7 + 2 \times 6 - 5 \times 3$           f) $2 \times 7 \times 8 - 3 \times 5 \times 6$

**Act 5  Scene 3  Mixed Operations With Fractions**

(Just as in Scene 1 of this Act, we always give the students products of fractions and natural numbers, with the denominator of the fraction a divisor of the natural number. The results will always come out as a natural number, and the calculations remain in the range of the students' abilities.)

Mentor: Now we return to expressions such as the ones in Act 5 Scene 1. Find the value of the expression: $\frac{2}{3} \times 9 + \frac{3}{4} \times 8$. George.

George: $\frac{2}{3} \times 9 = 6$ and $\frac{3}{4} \times 8 = 12$.

Mentor: Just a minute, George. How did you calculate $\frac{3}{4} \times 8 = 12$?

George: Oops! I multiplied 3 × 4, instead of 3 × 2. $\frac{1}{4} \times 8 = 2$, not 4!

Mentor: So, what should you get for the original expression?

George: $\frac{2}{3} \times 9 + \frac{3}{4} \times 8 = 6 + 6$, or 12.

Mentor: Now you are moving forward. Excellent.

Mentor: We will now evaluate $\frac{3}{2} \times 10 - \frac{2}{3} \times 6$. Jane.

Jane: I know that $\frac{3}{2} \times 10 = 15$, as half of 10 equals 5. Also, $\frac{2}{3} \times 6 = 4$, as $\frac{1}{3} \times 6 = 2$. So now I subtract 15 – 4 = 11.

Mentor: You have got the message. One for you, Alexander! $\frac{3}{5} \times 15 - \frac{1}{4} \times 12$.

Alexander: $\frac{3}{5} \times 15 = 3 \times 3$, or 9 and $\frac{1}{4} \times 12 = 3$, so we have 9 – 3 = 6.

Mentor: You are right on! Now let us try evaluating $\frac{7}{3} \times 6 + \frac{5}{6} \times 12$. Sarah.

Sarah: $\frac{7}{3} \times 6 = 7 \times 2$, or 14. Also, $\frac{5}{6} \times 12 = 5 \times 2$, or 10. So we have 14 + 10 = 24.

Mentor: Good work!

**Between the Scenes**

Simplify: a) $\frac{4}{5} \times 15 - \frac{3}{7} \times 14$   b) $\frac{3}{2} \times 16 + \frac{5}{3} \times 18$   c) $\frac{7}{4} \times 28 - \frac{5}{6} \times 12$

**Act 5**  **Scene 4**  **Yet Another Multiplication?**

Mentor: We have learned several ways of multiplying two numbers already, but this new method only requires that you can double and add numbers, giving you an alternative process you might use to check earlier calculations. Let's multiply $72 \times 39$ by the doubling-and-adding process. To do it, we set up two columns: the first for 72 and the second for 39, starting the latter with 1.

| 72 | 1 |
|---|---|
|  |  |
|  |  |
|  |  |
|  |  |
|  |  |
|  |  |

The second step requires that we double both 72 and 1, giving us 144 and 2, respectively. These numbers we enter in the second row.

| 72 | 1 |
|---|---|
| 144 | 2 |
|  |  |
|  |  |
|  |  |
|  |  |
|  |  |

We repeat this process until the sum of carefully chosen entries in the right-hand column equals 39. Can some one tell us what belongs in the third row?
Alexander: Yes, 288 and 4.
Mentor: George, give us the fourth row.
George: 576 and 8.
(The Mentor asks different students to fill in the rows up to but not including 64, being larger than 39.)

| 72 | 1 |
|---|---|
| 144 | 2 |
| 288 | 4 |
| 576 | 8 |
| 1152 | 16 |
| 2304 | 32 |
|  |  |

Mentor: Now comes the trickiest part. We want to mark each of the numbers in the right hand column that we need to get a sum of 39. We start with the largest number, 32. If we add 32 and 16 we get 48, a number bigger than our 39. So we will cross out the row of 16.

| 72 | 1 | √ | 39 |
|---|---|---|---|
| 144 | 2 | √ | 38 |
| 288 | 4 | √ | 36 |
| ~~576~~ | ~~8~~ | | |
| ~~1152~~ | ~~16~~ | | |
| 2304 | 32 | √ | 32 |
| 2808 | | | |

Mentor: What do we do with 8?
Sarah: If we add 32 and 8 we have a number, 40, also larger than 39. So we cross out 8 and 576.
Mentor: Excellent. What do we do with 4? Keep it or cross it out?
Tom: We can keep 4, because 32 + 4 = 36, a number less than 39.
Mentor: Fine. Put a check by 4 and move on to 2.
Jane: I know that 32 + 4 + 2 = 38, a number one less than 39, so we keep the 2.
Mentor: Good. We still do not have a sum of 39, but adding 38 + 1 will give us the result we want. Check 1! Now we have to add up all the checked numbers in the first column. So add: 72 + 144 + 288 + 2304. Alexander.
Alexander: 2808.
Mentor: How do we know 2808 is correct?
Alexander: I used lattice multiplication and got the same answer.
Mentor: Excellent. Use a different method, so you don't make the same mistake twice.

**Between the Scenes**
1. Multiply 96 × 897 by the lattice and the doubling methods. Which number should be over the 1's column in the doubling method? Why?
2. Multiply 3478 × 59 by both methods: lattice and doubling.

**Act 5**     **Scene 5**     **Consecutive Numbers**

Mentor: We speak of consecutive numbers when the numbers follow immediately on each other.  For example, 3 and 4 are consecutive numbers, while 3 and 5 are consecutive odd numbers, and 4 and 6 are consecutive even numbers.  Another example would be three consecutive counting numbers: 3, 4, 5.  George, name three consecutive numbers.
George: 1, 2, 3.
Mentor: Name three consecutive even numbers.  Sarah.
Sarah: 8, 10, 12.
Mentor: Name two consecutive odd numbers.  Alexander.
Alexander: 11 and 13.
Mentor: I want each of you to multiply two consecutive even numbers.  Tom, tell us what numbers you used and what result you got.
Tom: I multiplied 8 times 10 and got 80.
Mentor: Mary.
Mary: I multiplied 6 times 8 and got 48.
Mentor: Alexander.
Alexander: I multiplied 12 times 14 and got 168.
Mentor: I multiplied 4 times 6 and got 24.  Now I want each of you to add one to your product.  Doing that, I get $4 \times 6 + 1 = 25$.  George.
George: I multiplied 2 times 4 and added 1 to get 9.
Sarah: I calculated $10 \times 12 + 1 = 121$.
Alexander: $11 \times 13 + 1 = 144$.
Tom: $8 \times 10 + 1 = 81$.
Mary: $6 \times 8 + 1 = 49$.
Mentor: Do you see any pattern developing?
Mary: Yes.  We always get squares: 25, 9, 121, 144, 81, and 49.
Mentor: Good observation.  What can you say about the square roots of each of these numbers?
Mary: The square root of $49 = 7$, and seven is the odd number between my two numbers: 6 and 8.
Mentor: Tom.
Tom: The same here.  9 is the square root of 81 and lies between 8 and 10.
Sarah: 121 is 11 squared, and eleven is between my consecutive even numbers: 10 and 12.
Alexander: $12 \times 14 + 1 = 169$, but I don't know the square root of 169.
Mentor: Would you like to guess from what you have seen?
Alexander: How about 13?
Mentor: Show us how to square 13, using only doubling and addition.  Alexander.
Alexander: I make two columns: one for doubling 13, the other for doubling 1.

| 13 | 1 |
|----|---|
| 26 | 2 |
| 52 | 4 |
| 104 | 8 |
|  |  |

Now I start adding at the bottom of the 1 column until I reach a sum of 13. $8 + 4 = 12$. Adding 2 will go over 13, so I discard it and add just 1.

| 13 | 1 | √ |
|----|---|---|
| ~~26~~ | ~~2~~ |  |
| 52 | 4 | √ |
| 104 | 8 | √ |
| 169 | 13 | SUMS |

The corresponding numbers in the first column add up to 169. So $13^2 = 169$.
Mentor: I want you all to remember the first 15 squares, so let's calculate $15^2$. Again let's use the doubling-and-adding system.
Jane: We make two columns: one for 15, the other for 1.

| 15 | 1 | √ |
|----|---|---|
| 30 | 2 | √ |
| 60 | 4 | √ |
| 120 | 8 | √ |
| 225 | 15 |  |

We have to add all the numbers in the second column to reach 15, so we add all the numbers in the first column to reach 225.
Mentor: Jane, what do you think the value of $14 \times 16 + 1$ will equal?
Jane: $15^2$, because 15 lies between 14 and 16.
Mentor: You have the pattern! One last square will complete our collection: $14^2$. Tom.
Tom: We will make two columns again, one for 14, the other for 1.

| 14 | 1 |
|----|---|
| 28 | 2 |
| 56 | 4 |
| 112 | 8 |
|  |  |

This time we do not include 1 in our sum on the right and 14 on the left.

| ~~14~~ | ~~1~~ |   |
|--------|-------|---|
| 28     | 2     | √ |
| 56     | 4     | √ |
| 112    | 8     | √ |
| 196    | 14    |   |

Mentor: What observation can you make about the last two digits of $13^2$ and $14^2$?
Sarah: 96 and 69 are mirror images of each other!
Mentor: Did you learn something new? Have some fun?

**Between the Scenes**
Calculate the following squares: $16^2$, $17^2$, $18^2$, $19^2$, and $20^2$.

**Act 5      Scene 6      Exponents**

Mentor: In learning to multiply by doubling and adding, we have seen numbers like 16 over and over. We arrived at 16 by multiplying $2 \times 2 \times 2 \times 2$. If we count the number of twos, we find 4 of them and have a mathematical expression to so indicate: $2^4$. We read the expression as "two to the fourth" to indicate we are multiplying four twos together. We call 4 the **exponent**, 2 the **base,** and the expression an **exponential expression**. How do we arrive at the number 32 in this form of multiplication? Tom.
Tom: We multiply 5 twos together: $2 \times 2 \times 2 \times 2 \times 2$.
Mentor: How do we express this idea as an exponential expression?
Tom: 2 to the fifth or $2^5$.

Mentor: What is the value of $3^4$? Sarah.
Sarah: $3 \times 3 \times 3 \times 3 = 81$.
Mentor: What is the exponential expression for $5 \times 5 \times 5$, and what is its value?
George: $5 \times 5 \times 5 = 5^3$ or 125.
Mentor: What is the value of $4^3$? Mary.
Mary: $4^3 = 4 \times 4 \times 4$ or $16 \times 4 = 64$.

Mentor: Try to simplify $2^4 \times 2^5$. How many factors of 2 will we have?
Alice: $2^4 \times 2^5 = (2 \times 2 \times 2 \times 2) \times (2 \times 2 \times 2 \times 2 \times 2)$ or $2^9$.
Mentor: Can you guess the rule covering this product with the same base values?
Alice: We just add the exponents when multiplying two exponential expressions with the same bases.
Mentor: Retaining the bases 2 and 3 when we remove the parentheses, what form does the expression $(2 \times 3)^5$ take?
Mary: I would write: $2 \times 3 \times 2 \times 3 \times 2 \times 3 \times 2 \times 3 \times 2 \times 3 = 2^5 \times 3^5$. The exponent for the quantity becomes the exponent for each factor.
Mentor: Well said. You have discovered a lot about exponents.

**Between the Scenes**
1.   How many factors of three will you find in $9^5$? In $27^4$?
2.   How many factors of two will you find in $4^7$? In $12^7$? In $8^6$?
3.   Evaluate: $2^5 \times 3^4$.
4.   Evaluate: $2^3 \times 5^4 \times 3$.

**Act 5**     **Scene 7**     **Mixed Operations With Exponents**

Mentor: We have already agreed to take multiplication and division before addition and subtraction. So 3 × 5 – 2 × 4 = 15 – 8 or 7. However, working strictly from left to right, multiplying 3 × 5, then subtracting the 2 from the 15, and finally multiplying 13 times the 4, gives us the incorrect 15 – 2 = 13 and 13 × 4 = 52. Some programming languages always calculate strictly from left to right, illustrating that the **order of operations as a mutual agreement** rather than a necessary property. So where does exponentiation fit into this business? We will agree to make it come before multiplication and division, and therefore before addition and subtraction. Let us calculate $2 \times 3^4 + 5$.

$$2 \times 3^4 + 5 = 2 \times 81 + 5, \text{ and}$$
$$2 \times 81 + 5 = 162 + 5, \text{ and}$$
$$162 + 5 = 167$$

Calculate: $4 \times 5^4 - 2 \times 6$. Alexander.

Alexander: $4 \times 5^4 - 2 \times 6 = 4 \times 625 - 12$, and $4 \times 625 - 12 = 2500 - 12$, or 2488.
Mentor: Exactly. How about calculating $3 \times 2^5 + 7$? Jane.
Jane: $3 \times 2^5 + 7 = 3 \times 32 + 7$, and $3 \times 32 + 7 = 96 + 7$, or 103.
Mentor: Good. We can summarize our mixed operations with the following table, a table expressing the first operation at the top and the last at the bottom. The exponentiation symbol frequently used in computers is the caret: ^. We call our table a **precedence** table.

| ^ | |
|---|---|
| × | / |
| + | - |

Mentor: We want to be comfortable evaluating expressions with mixed operations, so you should ask questions, if you are confused. Evaluate: $400 - 2 \times 5^3$.
Alice: $400 - 2 \times 125 = 400 - 250$ or 150. No questions!
Mentor: Well done.

Did you learn something new? Have a little fun?

**Between the Scenes**
1. Calculate: $5 \times 4^3 - 3 \times 2^4$.
2. Calculate: $11^2 + 5 \times 3^4$.
3. Evaluate: $3 \times 5^3 - 2 \times 7^2$.

**Act 5**  **Scene 8**  **Changing Precedence**

Mentor: No sooner than we tell you what order we will calculate expressions with the five operations, we show you how to change that order. Suppose you wanted to multiply a number times the sum of two other numbers, such a 5 multiplied times the sum of 3 and 8. We do not want to multiply the 5 times the 3 and then add 8, which is what we get if we write: $5 \times 3 + 8$. How can we indicate that the addition should come first? We use **parentheses**: "(" and ")". Whatever appears inside the pairs of parentheses must be done first, before whatever the operation outside indicates. To multiply 5 times the sum of 3 and 8, we write: $5 \times (3 + 8)$ and say "five times the quantity three plus eight." We also say that 3 and 8 are **summands**, because we "sum" them. Calculating, we find:
$$5 \times (3 + 8) = 5 \times 11, \text{ and}$$
$$5 \times 11 = 55.$$

Now calculate: $3 \times 7 + 11$ and $3 \times (7 + 11)$. Fred, do the first one; George, the second.

Fred: $3 \times 7 + 11 = 21 + 11$, or 32.

George: $3 \times (7 + 11) = 3 \times 18$, and $3 \times 18 = 54$.

Mentor: Exactly. These two examples show us the importance of the order of operations. Let's mix exponentiation with subtraction: just the two operations.

Simplify  a) $2^5 - 3$ and
  b) $2^{(5-3)}$

Jane, tell us what to do.

Jane: To simplify the first expression, $2^5 - 3$, with no parentheses, we have to raise 2 to the fifth first and then subtract 3. We have $2^5 = 32$, and $32 - 3 = 29$.

To simplify the second expression, $2^{(5-3)}$, we have to subtract 3 from 5 first and then raise 2 to that power. Here we have $2^{(5-3)} = 2^2$, and $2^2 = 4$.

Of course, 29 does not equal 4.

Mentor: Yes. Good analysis. Next we will try multiplication and subtraction:
$$2 \times 7 - 4 \text{ and } 2 \times (7 - 4).$$

Sarah: $2 \times 7 - 4 = 14 - 4$ and $14 - 4 = 10$. We have to multiply before we subtract.

$2 \times (7 - 4) = 2 \times 3$, because we have to perform the operation inside the parentheses first: $7 - 4 = 3$. Then we multiply by 2 to get 6.

Mentor: Next we mix subtraction and division. With $6 - 4/2$ and $(6 - 4)/2$.

Tom: In the first expression, we will divide 4 by 2 before we subtract that from 6.

So we have $4/2 = 2$, and $6 - 2 = 4$.

But in $(6 - 4)/2$ we subtract first, $6 - 4 = 2$, and then divide 2 by 2 to get 1.

Mentor: Let's mix addition and exponentiation: $5 + 7^2$ and $(5 + 7)^2$. Mary.

Mary: With $5 + 7^2$, we must square 7 first and then add 5. So we have $5 + 49 = 54$.

With the second expression, $(5 + 7)^2$, we add 5 and 7 first for 12 and then square that for 144.

Mentor: Now try a more complicated exponential expression: $7 \times 3^5 - 2 \times 11$.

Alice: $7 \times 3^5 - 2 \times 11 = 7 \times 243 - 22$, and $7 \times 243 - 22 = 1701 - 22$, or 1679.

Mentor: Nice work. Try: $6 \times (7^2 - 13)$.

Alice: $6 \times (7^2 - 13) = 6 \times (49 - 13)$, or $6 \times 36$, and $6 \times 36 = 216$.

Mentor: That does it! Let's try one more. Evaluate: $2 \times (190 - 4 \times 2^5)$. Tell us the order of the operations.

Sarah: Raise two to the fifth,
      Multiply that by 4,
      Subtract that from 190, and then
      Multiply by 2.

Mentor: An excellent indication of the order of operations. What is the value?

Sarah: $2 \times (190 - 4 \times 32) = 2 \times (190 - 128)$, and
     $2 \times (190 - 128) = 2 \times 62$, or 124.

Mentor: We will need to practice this often as the order of operations forces us to calculate everything in a certain sequence. We will find ourselves using these orders frequently as we progress in mathematics.

**Between the Scenes**

1.     Without calculating the value, indicate the order of the operations in the following expression: $29 \times (15 \times 3^4 - 13)$.
2.     What is the order of operations in the following: $9 \times 5 + 17$ and $9 \times (5 + 17)$?
3.     Evaluate: $6 \times 2^{(3+4)}$, indicating the order of operations.
4.     Evaluate: $(5 \times 2)^7$, indicating the order of operations.

**Act 6     Prime Numbers, GCD and LCM**

**Scene 1     Prime Numbers Reexamined**

Mentor: We have seen how using a prime or a composite number can change the nature of our calculations. In clock arithmetic, multiplying four times six gave us zero even though the four and six did not equal zero themselves. In modular arithmetic based on seven, the only products equal to zero had one of their factors equal to zero. In multiplying fractions and natural numbers, we had to be sure the denominator of the fraction was a factor of the natural number, or the result of our computation would equal a second fraction, not the natural number we expected. So let us review the idea of a prime number. Who can remember how we knew a prime number? Tom.
Tom: If we could only find two one-color trains equal to our given train, then the given train belongs to the prime numbers.
Mentor: Perfect. What did we do to show that yellow (or 5) is prime? What trains did we consider?
Tom: Red, green, and purple trains.
Mentor: What about white and yellow trains?
Tom: White and the given color always work, so we do not have to worry about them.

Mentor: Let's consider the three colors again. What happened to the red trains we tried? Sarah.
Sarah: Two red rods fell short by a white rod, and three reds reached beyond the yellow train by one white.
Mentor: Good. Consider the green trains. Alexander.
Alexander: One green rod fell short and two fell long.
Mentor: Does your experience with the green trains suggest what will happen with the purple trains?
Alexander: Yes. A green rod is so long a second one would reach beyond the yellow, so purple will fail in the same way.
Mentor: Ah! So, if we want to show that 11 is a prime number, what numbers, or colors, will we have to check out and which ones not? Jane.
Jane: I think we will have to check all of the numbers less than 11, so red through orange.

Mentor: Alexander, you disagree?
Alexander: Yes. For red through yellow (2 through 5) we will have to build trains, but one dark green rod is too short and two are too long. So dark green is not a factor of eleven. The same will hold for black through orange (7 through 10).
Mentor: Good argument. Think about a rule we might make that will tell us when we can stop checking any more trains. For now let's look at the shorter rods. Red, Mary.
Mary: Five red rods makes a train falling one white rod short, while a train of six red rods comes out beyond 11.
Mentor: What do we find with green rods? George.

George: A train of three green rods falls short by a red rod and one of four extends too far by a white one.
Mentor: Purple? Henry.
Henry: A train of two purple rods fails to reach to eleven, and one of three falls beyond eleven.
Mentor: Looking at the yellow rods, what do you see? Tom.
Tom: A train of two is too short and of three is too long.

Mentor: So we now know 11 is prime. But we want to go back to consider the most economical way to determine if any number is prime. Can we name certain numbers we will not have to examine? This will cut down on the amount of work we will have to do to be certain a number is prime. Which trains will we not have to consider? Tom.
Tom: If two of the rods of one color extend too far, then all the longer rods will do the same.
Mentor: Is 12 prime? Mary.
Mary: No. 2 is a factor, so a red train will do. This makes the third one-color train equal to 12.
Mentor: What colors will we not have to consider in order to determine if the number 13 is prime? Sarah.
Sarah: Black trains and any longer rods will miss the mark. 7 equals half of 14 and so is more than half of 13.

**Between the Scenes**
What potential factors must we check to be sure the number 17 is prime? We know we will have to start with 2 and 3 as the smallest possible factors. Starting with the largest numbers less than 17, would you try 16 or 15? Why or why not? What is the largest factor candidate you think we would have to try? Why?

**Act 6      Scene 2           One-color to One-class Trains**

Mentor: Now we are discussing numbers bigger than ten, so we will have to change our definition of prime numbers. We do not have a single rod the length of 11. We have to use two rods; usually an orange and a white, but black and purple would also give us the correct length. We will refer to a **one-class** train, if all the components have the same color combinations. We can make a one-class train of length 22 with two trains, each made up of an orange and a white rod. So 22 is not prime, but composite, because we have a one-color red train of equal length, but also a one-class orange-white train. Every one-color train we will think of as being a one-class train. So we call a number **prime** if it has exactly two one-class trains of the same length—the easy one being a white train, the other the unique given train.

      Going back to 13, what one-class trains must we try to show it is prime? Sarah.
Sarah: We will have to try all the one-class trains shorter than black, or 7. We know that 13 is not even, so none of the red group rods (red, purple, dark green, brown, orange) will make a train of equal length. Otherwise 13 would be even.
Mentor: Nice. You have just thrown out five trains we will not have to look at. What colors must we consider? Tom.
Tom: Green and yellow, or 3 and 5. 4 green rods make a train one white rod short, but 5 make a train one red rod too long. Yellow, or 5, works the same way: 2 yellows make a train three whites short while 3 makes a train two whites too long.

Mentor: Let us try one more number: 19. What numbers must we examine to be sure 19 is prime?
Jane: 10 is half of 20, so more than half of 19. We will have to look at all the numbers from red to blue, one less than orange.

| 2 | 3 | 4 | 5 | 6 | 7 | 8 | 9 | 10 |
|---|---|---|---|---|---|---|---|----|

19 is not even, the units digit is 9, so all the even numbers from 2 to 10 go out.

| ~~2~~ | 3 | 4 | 5 | ~~6~~ | 7 | 8 | 9 | ~~10~~ |
|---|---|---|---|---|---|---|---|----|

That leaves us: 3, 5, 7, and 9. 3 does not divide 19, because 6 × 3 is too short and 7 × 3 is too long. And 9 does not divide 19, because if it did 3 would divide 19.

| ~~2~~ | ~~3~~ | 4 | 5 | ~~6~~ | 7 | 8 | ~~9~~ | ~~10~~ |
|---|---|---|---|---|---|---|---|----|

19 does not end with a 5 or a 0, so 5 does not divide 19.

| ~~2~~ | ~~3~~ | 4 | ~~5~~ | 6 | 7 | 8 | 9 | ~~10~~ |

Lastly, 2 × 7 = 14, which is too short, and 3 × 7 = 21, which is too long, so 7 does not divide 19.

| ~~2~~ | ~~3~~ | 4 | ~~5~~ | 6 | ~~7~~ | 8 | 9 | ~~10~~ |

Mentor: Wow. You raced through that and didn't miss a thing!

Did everybody learn something new?   Yes. What?
Have some fun?   Yes. Why?

**Between the Scenes**

Can you demonstrate that 23 belongs to the collection of prime numbers?

**Act 6**  **Scene 3**  **Reduced Prime Test**

Mentor: We have seen in the last scene how we can argue that 19 belongs to the set of prime numbers by testing just those numbers half as large as 19 (or 20). So we checked all the natural numbers from 2 through 10. The improvement, using just the numbers less than half of 19, meant we only had to test 2 through 9 and not 10 through 18, roughly half our potential divisors. Let's repeat the experience with a larger number, say 37. What do we have to do to show 37 is prime? Jane.

Jane: Half of 38 equals 19, so we have to eliminate all potential factors from 2 through 19.

| 2 | 3 | 4 | 5 | 6 | 7 | 8 | 9 | 10 | 11 | 12 | 13 | 14 | 15 | 16 | 17 | 18 | 19 |
|---|---|---|---|---|---|---|---|----|----|----|----|----|----|----|----|----|----|

We know 37 is not even, because its unit's digit is 7. So we eliminate 2, and all the other even numbers.

Mentor: Why can we cross out all the other even numbers?

Jane: If 6 divided 37, then 2 would divide 37, but we know 2 does not divide 37. So 6 divides 37 contradicts the statement that 2 does not divide 37.

Mentor: Excellent. By finding that 2 does not divide 37, we managed to throw out nine numbers: 2 and all its multiples. What do we do next? Tom.

| ~~2~~ | 3 | ~~4~~ | 5 | ~~6~~ | 7 | ~~8~~ | 9 | ~~10~~ | 11 | ~~12~~ | 13 | ~~14~~ | 15 | ~~16~~ | 17 | ~~18~~ | 19 |
|---|---|---|---|---|---|---|---|---|---|---|---|---|---|---|---|---|---|

Tom: We look at 3, a prime number, and all its multiples: 6, 9, 12, 15, 18. When we divide 3 into 37, we get a remainder of 1, so it's not a divisor of 37. Now we can throw out the multiples of 3 that are odd, as we have already dropped the even multiples of 3.

| ~~2~~ | ~~3~~ | ~~4~~ | 5 | ~~6~~ | 7 | ~~8~~ | ~~9~~ | ~~10~~ | 11 | ~~12~~ | 13 | ~~14~~ | ~~15~~ | ~~16~~ | 17 | ~~18~~ | 19 |
|---|---|---|---|---|---|---|---|---|---|---|---|---|---|---|---|---|---|

This leaves us with a much smaller list: 5, 7, 11, 13, and 17.

Mentor: What comes next? Sarah.

Sarah: 5 does not divide 37, because 37 does not end in 0 or 5. We have already tossed out 10 and 15, so nothing else falls out.

Mentor: Now we have just 7, 11, 13, and 17 left. Tell us how to continue. Mary.

Mary: 7 does not divide 37, because $7 \times 5 = 35$, meaning that we will have a remainder of 2 when we divide by 7.

Mentor: Excellent. Next? Alexander.

Alexander: We just have 11, 13, and 17 to go. 3 time 11 equals 33, so we will have a remainder of 4 upon dividing 37 by 11. So 11 also goes out.

Mentor: Just two to go! George.

George: 2 times 13 equals 26, so if we divide 37 by 13 we will get a remainder of 11. Out goes 13!

Mentor: We know that two times seventeen equals 34, leaving us a remainder of 3. So now we have shown 37 is prime. Shifting to a new prime-checking pattern: Can we cut down on the list of potential factors to show a number is prime? Do we have to check up to half way? Consider the much smaller table of the factors of twelve.

| | |
|---|---|
| 1 | 12 |
| 2 | 6 |
| 3 | 4 |
| 4 | 3 |
| 6 | 2 |
| 12 | 1 |

Do you see a pattern here? Jane.

Jane: The bottom half looks like the top half seen in a mirror through the two threes.

Mentor: Beautiful. So what is the highest number we have to check for divisibility? Did we have to check out 6? Tom.

Tom: No. When we divided 12 by 2, we got the other factor 6.

Mentor: If we are looking for all the factors of twelve we will have found them after we have considered what number?

Jane: 3, and 3 is less than half of twelve.

Mentor: So if we wanted to draw a line between those factors of 12 we would have to consider in order to find all the factors of 12, where would we draw the line?

George: Between 3 and 4.

| | |
|---|---|
| 1 | 12 |
| 2 | 6 |
| 3 | 4 |
| Here | |
| 4 | 3 |
| 6 | 2 |
| 12 | 1 |

Mentor: How can we describe this location? Look at another example: 16.

| | |
|---|---|
| 1 | 16 |
| 2 | 8 |
| 4 | 4 |
| 8 | 2 |
| 16 | 1 |

Where should we place the camera lens to see the top in the bottom? Sarah.

Sarah: The lens would go between the two 4's in the middle of the line dividing the two cells.

Mentor: State the relationship between 4 and 16. Sarah.

Sarah: 4 squared equals 16 or the square root of 16 equals 4.
Mentor: Exactly. So for our demonstration that 37 is prime, where could we have drawn our line?
Sarah: At the square root of 37, between 6 and 7.
Mentor: If we had done that, what numbers must we have considered?
Sarah: 2, 3, 4, 5, 6.
Mentor: Wouldn't that have made the assignment easier?
Class: YES!
Sarah: Why didn't you tell us to do the test this way the first time?
Mentor: Several reasons come to mind immediately. First, you needed to understand the problem and the one-color train solution. Second, you needed to recognize that this problem, like most other problems, has a number of solutions. Third, you appreciate the square root solution much more because you have been through the other, less efficient solutions.
So let's tackle 97! Is 97 prime? Tom, start us off!
Tom: The square root of 100 is 10, so we only have to consider numbers less than 10.
Mentor: Excellent. Mary, what do we do next?

| ~~2~~ | 3 | 4 | 5 | 6 | 7 | 8 | 9 |
|---|---|---|---|---|---|---|---|

Mary: We see from the units digit, 7, that 97 is odd, so not divisible by 2.
Mentor: OK. George, what do we do next?
George: Cross out all the other even numbers in our list.
Mentor: Good, but what justifies doing that?
George: If 6, or some other even number, divided 97, then 2 would divide 97. But we know that 2 does not divide 97.

| ~~2~~ | 3 | ~~4~~ | 5 | ~~6~~ | 7 | ~~8~~ | 9 |
|---|---|---|---|---|---|---|---|

Mentor: Jane, tell us what to do now.
Jane: We ask if 3 divides 97. It does not because 3 divides 96, so will have a remainder of 1 going into 97. This means that 6 and 9 will also go.

| ~~2~~ | ~~3~~ | ~~4~~ | 5 | ~~6~~ | 7 | ~~8~~ | ~~9~~ |
|---|---|---|---|---|---|---|---|

Mentor: Alexander, can you help us clean up from here?
Alexander: Yes. 5 does not divide 97, because 5 only divides numbers with 5 and 0 in the units digit place. This just leaves us 7.

| ~~2~~ | ~~3~~ | ~~4~~ | ~~5~~ | ~~6~~ | 7 | ~~8~~ | ~~9~~ |
|---|---|---|---|---|---|---|---|

We know that 7 goes into 77 evenly, but adding 20, that 7 does not divide evenly, will give us a non-zero remainder.

| ~~2~~ | ~~3~~ | ~~4~~ | ~~5~~ | ~~6~~ | ~~7~~ | ~~8~~ | ~~9~~ |
|---|---|---|---|---|---|---|---|

Mary: Could I argue that 7 goes into 97 just 13 times with a remainder of 6, so 7 does not divide 97 evenly?

Mentor: Excellent. What are the numbers we really had to examine to come to our conclusion? Sarah.

Sarah: We had to look at the prime numbers less than 10, namely: 2, 3, 5, 7.

Mentor: What numbers must we test to discover if 137 is prime?

Jane: The square root of 144 equals 12, so we would have to test 2, 3, 5, 7, and 11. Those are all the primes less than the square root of 137.

Mentor: Look how few test numbers we have! Do any of these numbers divide 137?

Jane: No. 2, 5, and 11 are trivial. 3 goes into 137 45 times with a remainder of 2. 7 does not divide 137 either (7 × 19 = 133 and 7 × 20 = 140).

Mentor: We have reduced the problem as far as we can go. I think we have solved the problem. Did you learn something new? Have some fun?

**Between the Scenes**

1. <u>If a number divides one summand of a sum of two summands but does not divide the other summand, then the number does not divide the sum</u>, states a divisibility rule. Give an example where this rule applies. (Recall that we call the numbers 2, 3, and 4 in the sum 2 + 3 + 4 **summands**—numbers being summed.)

2. Apply the rule that 'the root of the product equals the product of the roots' to simplify a radical (root) expression having two factors, one a square, the other not a square. Example: Simplify $\sqrt{12} = \sqrt{4 \times 3}$, observing that $\sqrt{4} \times \sqrt{3} = 2\sqrt{3}$. Now there is no square, bigger than one, inside the radical. Simplify: $\sqrt{45}$ and $\sqrt{75}$.

3. Similarly, simplify each of the following: $\sqrt{\frac{8}{45}}$, $\sqrt{\frac{75}{49}}$, and $\sqrt{\frac{27}{8}}$. State the 'rule' you are using in each step.

4. Another 'rule' states: No denominator in the radicals and no radicals in the denominator. When we simplify $\sqrt{\frac{4}{5}}$, we can find the square root of 4 as 2, but are left with 5 in the denominator. So we have $2\sqrt{\frac{1}{5}}$. To avoid a denominator, here a 5, in the radical, we multiply both numerator and denominator by 5 to give us $\sqrt{\frac{4 \times 5}{5 \times 5}}$. Now we can simplify the expression to $\frac{2}{5}\sqrt{5}$, leaving no fraction inside the radical. Simplify the following so no fraction remains inside the radical: $\sqrt{\frac{8}{3}}$, $\sqrt{\frac{98}{27}}$, $\sqrt{\frac{28}{125}}$.

**Act 6        Scene 4              Least Common Multiple**

Mentor: When we do our calculations, we will try to work with the smallest numbers possible, so we can do much of the work in our heads and see possible errors quickly. Finding the least common multiple, LCM, of two numbers has its most prevalent application in adding and subtracting fractions, a topic we will deal with soon.
Problem: Find the least common multiple of 6 and 9.
Solution: Start with two parallel one-color trains: one dark green, the other blue.

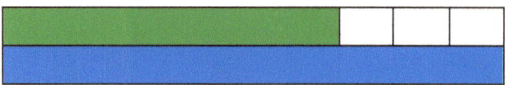

What do we have to do to make the trains the same length?
Jane: Make the dark green train longer. Add one more dark green.

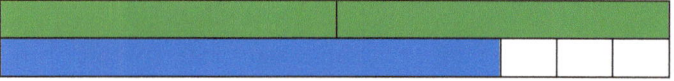

Mentor: Now the dark green train extends too far. What do we do?
Jane: Add another blue rod to the blue train.

And when we add another dark green, we will find the two trains are the same length, 18.
Mentor: Correct. 18 is a multiple of both 6 and 9, and it is the first one we arrive at by building trains. In this case, we see that the LCM is smaller than the product of the two numbers, which is 54. Will this always be the case?
Mary: I think 5 and 7 will have the same LCM and product: 35.
Mentor: Why do you think this?
Mary: Both are prime numbers, so we will need seven yellow rods and five black rods before the two trains are equal.
Mentor: Nice reasoning. Try 4 and 9. What will their LCM be?
Mary: Oh, I see. I will have to have nine 4's and four 9's to create equal trains, but 4 and 9 are not prime.
Mentor: Exactly. What number divides both 4 and 9?
Mary: Just 1.
Mentor: When 1 is the largest common factor of two numbers, we say the numbers are **relatively prime**. This brings us to our next topic: Greatest Common Divisors or GCD's. Did you learn something new? Have some fun doing it?

**Act 6**   **Scene 5**   **Euclidean Algorithm & GCD**

Mentor: In the last scene, we have used trains of rods to find the least common multiple, **LCM**, of two numbers. In this scene we take up a related concept, that of the greatest common divisor or **GCD**. Most of you can tell us the GCD of 6 and 15.

Alexander: Both 1 and 3 are divisors of 6 and 15. 3 is the largest of the divisors.

Mentor: How did you know 3 was GCD so quickly?

Alexander: I can factor 6 into $2 \times 3$ and 15 into $3 \times 5$, seeing that 3 is the only common factor other than 1.

Mentor: Let us consider the GCD of two numbers with factors we might not recognize: 143 and 221. Let's start off asking how many times the smaller number goes into the larger.

George: I see that 2 times 143 will be larger than 221, so the answer is $1 \times 143$ with a remainder of 78.

Mentor: Excellent. We will write your observation in a traditional form called the **Euclidean Algorithm**, but don't let the name scare you!

$$221 = \underline{143} \times 1 + \mathbf{78}$$

Now to find the GCD we repeat this exercise by sliding the 221 out of the equation and replacing it by the 143, and replacing the original 143 by the remainder, 78.

$$\underline{143} = \mathbf{78} \times 1 + 65$$

We continue sliding and replacing until we get a remainder of 0.

George: This gives me

$$221 = \underline{143} \times 1 + \mathbf{78},$$
$$\underline{143} = \mathbf{78} \times 1 + 65,$$
$$\mathbf{78} = 65 \times 1 + 13,$$
$$65 = 13 \times 5 + 0$$

Mentor: Now I claim that the last non-zero remainder, 13, is the GCD of 221 and 143.

George: $13 \times 11 = 143$ and $78 = 6 \times 13$, so 13 divides $143 + 78$ or 221!

Mentor: Good! You have completed the process for two larger numbers, numbers with factors you might not recognize immediately. If you recognize all the factors, the GCD is the largest common factor, but for large unfamiliar numbers the Euclidean Algorithm will give you the desired result without guessing.

  Now calculate the product of the GCD, 13, and the LCM, 2431, and compare that to the product of the two original numbers, 143 and 221. What do you observe about these two products? What pattern might this illustrate?

**Between the Scenes**
Find the GCD and LCM of:   (1) 77 and 121   (2) 156 and 182   (3) 48 and 75
(4) 204 and 323.

**Act 7**  **Areas and Parentheses**

**Scene 1**  **Rectangles**

Mentor: We describe a **rectangle** as a four-sided shape with straight sides, the adjacent sides being perpendicular to each other. All of the faces of our Cuisenaire Rods form rectangles, all of the faces of the white rods having equal edges, forming **squares**. The squares measure one centimeter on each side. A meter measures one hundred centimeters, looked upon as the basic unit of the **metric** system, as opposed to the **English** system, consisting of feet, yards, miles, etc. Currently, our schools introduce the metric system while American society measures most everything with the English system.

If we take six white rods and line them up in two rows of three each, we have the rectangle on the left.

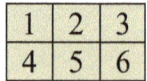 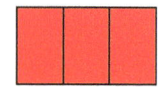 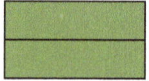

We say that this rectangle has a width of 2 cm, a length of 3 cm, and an area of $2 \times 3$ or 6 $cm^2$. We might have made this rectangle with three red rods (in the middle) or two green rods (on the right), illustrating that $3 \times 2 = 2 \times 3$, the **symmetric property of multiplication**. But, perhaps, the easiest way to think about the area is to count the number of white rods filling the shape. Notice that each of the smallest rectangles is a square, 1 centimeter (1 cm) on each side. Jane, show us a rectangle 4 cm by 6 cm.
Jane: Here is a 4-by-6 rectangle.

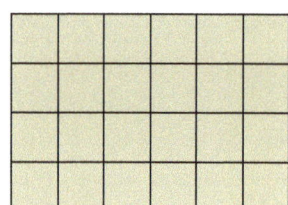

Mentor: What do we call two even numbers (4 and 6) coming one after another? And how many square centimeters in the area?
Jane: 4 and 6 are **consecutive** even integers. And the area of the 4-by-6 rectangle shown above equals 24 square centimeters, the product of the consecutive even integers.
Mentor: Now, Jane, I want you to take the four rods in the sixth column off the rectangle on the next page and place them in a new fifth row on the top.

Jane: Here. I have taken the four squares in the last column and put them on top of the rectangle as a top partial row in purple.

Mentor: Have you got a rectangle?
Jane: No. But if I added one white rod in the upper right-hand corner, I would have a square!
Mentor: How many white rods do you have in the square?
Jane: 25.
Mentor: What do we know about the square root of 25?
Jane: The square root of 25 equals 5, and 5 is between the consecutive even numbers 4 and 6. We observed this earlier, back in Act 5 Scene 5!
Mentor: Back then we observed that 4 × 6 + 1 = 25. Suppose I had picked two other consecutive even numbers: 6 and 8. How would you show with the rods that our earlier rule applies?
Jane: Whenever we take the last column, shown here in green, off the long side of the 6-by-8 rectangle and put it on top as a row, we subtract one column, 8 – 1 = 7, and add one row, 6 + 1 = 7, except at the corner. The area of the figure formed will always fall one white rod short of the square of the number between the consecutive even integers. So here we have 6 × 8 + 1 = $7^2$.

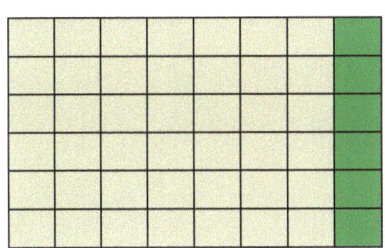 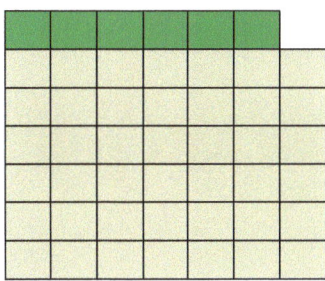

Mentor: Excellent observation. Now we can see a general rule: **For consecutive even integers, taking one column off the end of the longer rows, turning that column into a new row, and adding one white rod, forms a square with area equal to the square of the odd number between the given even numbers.**

Now we are going to put the 1 on the side with the 49, so that we have a difference: 6 × 8 = 49 – 1. You know that 1 equals the square of 1, so we can call the expression on the right side of the equation the **difference of two squares**. We will use this pattern time and again and develop the thought in the next Scene.

**Act 7      Scene 2           The Difference of Two Squares**

Mentor: The difference of two squares does not have to have the smaller square be a one-by-one, but could be any other sized square. The way we will look at the difference of two squares will always have the smaller square at one corner of the larger. This configuration will allow us to reverse the process we have just been through: calculating the area of the difference of two squares by cutting off, say, a column, and attaching it to the bottom of the rectangle that is left to make a larger rectangle. Here we have the sequence for $9^2 - 5^2$.

We do not know a formula for the area of the blue region, an L-shape. So we will cut along the vertical line between the yellow and blue up to the top of the larger square. This leaves us two blue rectangles: one 9 by 4, the other 5 by 4. We can rotate the smaller rectangle, 5 by 4 below, and attach it to the larger rectangle using the common sides, each of length 4.

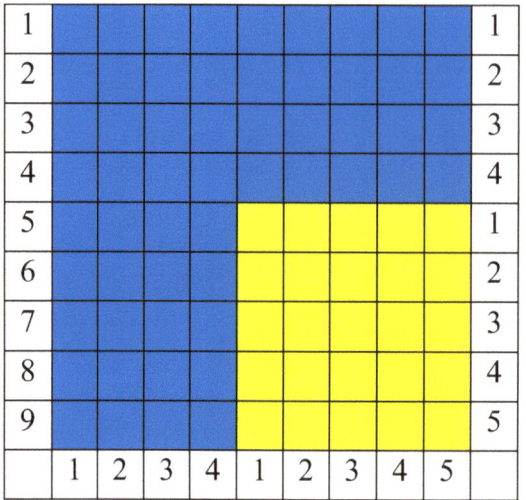

Our new rectangle has 14 white rods down and 4 across, so we know the area equals 56. We also know that $56 = 81 - 25$, the difference of two squares. Looking a little closer, we see a pattern requiring the use of parentheses;
$81 - 25 = (9 - 5) \times (9 + 5)$, or $4 \times 14$.

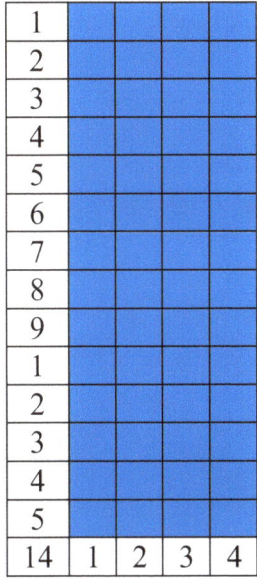

The 9 – 5, or 4, constitutes the width across the bottom; the 9 + 5, or 14, the height or length from top to bottom. We have converted an L-shaped figure without a formula for its area to a rectangle with an easy formula for its area: L × W = Area.

Let's try another difference of two squares: $8^2 - 3^2$. Tom.

Tom: I will make a square 8 on each side with a smaller square 3 on each side cut out of the lower right corner.

Tom continues: Next I will cut off the top right brown section, following the vertical line between the two colors. The longer piece makes a rectangle 8 by 5, while the smaller piece a rectangle 5 by 3. We can rotate the smaller piece around and attach it to the bottom of the longer piece, giving us a new rectangle 11 by 5, with an area of 55 square centimeters.

Mentor: That describes the process well. What did we start with?

Tom: An 8 by 8 square minus a 3 by 3 square.

Mentor: What does the equation look like?
Tom: $8^2 - 3^2 = (8 - 3) \times (8 + 3)$.
Mentor: Before we move on to other topics, we should observe that the equations we see in this scene equate a <u>difference</u> with a <u>product</u>:

$$81 - 25 = (9 - 5) \times (9 + 5), \text{ or}$$
$$9^2 - 5^2 = (9 - 5) \times (9 + 5),$$

in the first example and

$$64 - 9 = (8 - 3) \times (8 + 3), \text{ or}$$
$$8^2 - 3^2 = (8 - 3) \times (8 + 3)$$

in the second. This gives us a formula we will use many times in this course and in the future. <u>The difference of the squares of two numbers equals the product of the difference of the two numbers with the sum of the two numbers.</u> We call changing a difference into a product in this manner **factoring**. We can find similar patterns in other expressions, if we just look for them. We should also note here the important role the parentheses play, making subtraction and addition come before multiplication.

**Between the Scenes**
1. Calculate $456^2 - 454^2$ two ways: by squaring and by factoring. Which way is simplest?
2. Factor: $11^2 - 8^2$.
3. Observing that "no radicals in the denominator and no denominators in the radicals" characterizes a simplified expression, we could simplify $\frac{1}{\sqrt{2}}$ by multiplying the numerator and denominator by $\sqrt{2}$. This gives us the expression $\frac{1\sqrt{2}}{\sqrt{2}\sqrt{2}} = \frac{\sqrt{2}}{2}$.

Now simplify: $\frac{2}{\sqrt{3}}$.

4. The difference of two squares gives us a way to simplify expressions with sums or differences in the denominators. Simplify: $\frac{1}{\sqrt{5} - \sqrt{3}}$.

Solution: $\frac{1}{\sqrt{5} - \sqrt{3}} = \frac{\sqrt{5} + \sqrt{3}}{(\sqrt{5} - \sqrt{3})(\sqrt{5} + \sqrt{3})}$, or $\frac{\sqrt{5} + \sqrt{3}}{5 - 3} = \frac{\sqrt{5} + \sqrt{3}}{2}$.

Now simplify: $\frac{1}{\sqrt{7} + \sqrt{2}}$.

**Act 7      Scene 3      Ross's Rule**

Mentor: Ross served in the army during the Second World War, having as one of his duties to blow up trees blocking the unit's passage. The problem he faced in these situations required that he square some measurement of the tree, such as its radius, diameter or circumference. The army had trained him to square two-digit numbers ending with 5, such as 75, and Ross rounded off his measurements, such as 73, to the nearest multiple of 5, here 75.  The rule told him to take the tens digit, 7, and multiply it by the tens digit plus one, 8, for $7 \times 8 = 56$, and then tack 25 on the end, combined to give him 5625. This formula works! You can impress your friends and neighbors by showing them your speed of computation. What does the square of 35 equal? Jane.

Jane: The square ends in 25. So I multiply $3 \times (3 + 1)$ to get 12 and tack on 25, arriving at 1225.

Mentor: Nice.  Later, we will return to explain why this works, but for now let's calculate $65^2$. Alexander.

Alexander: 6 times 7 equals 42, with 25 put on the end, gives us 4225.

Mentor: Of course, if the measurements came out close to a number ending with zero, then the computation was trivial ($60^2 = 3600$): just square the tens digit and add two zeros.

Now you know how to use Ross' Rule!

**Between the Scenes**

1.      Use lattice multiplication to show why Ross's rule works. Specifically, why are the last two digits always 25? Note: if I square other numbers, say 46, the last two digits are not 36 or $6 \times 6$.

2.      Use lattice multiplication to find a two-digit number, without 0 or 5 as the unit's digit, such that the last two digits of the square of the number equal the square of the last digit of the original number. Why does your number work? Note: the square of 46 equals 2116, not 2136. The square of 44 equals 1936, not 1916.

3.      Can you find more than one two-digit number, without 0 or 5 as the unit's digit, such that the last two digits of the square form the square of the unit's digit of the original number?

**Act 7**  **Scene 4**  **Adding Areas for New Formulas**

Mentor: We are going to look into the use of parentheses again now. Best think about the areas of one or more rectangular regions, so when we multiply the length times the width we can picture all the little white rods filling the space. Once again, you want to look for the patterns.

My neighbor and I like to garden. So, when I hire the farmer to till my garden, I have the farmer till my neighbor's garden also. To make the job easier, we have arranged our gardens next to each other. My plot forms a rectangle 7 meters by 5 meters; my neighbor's a plot 9 meters by 5 meters. My garden, on the left, I colored green; my neighbor's, on the right, he colored red.

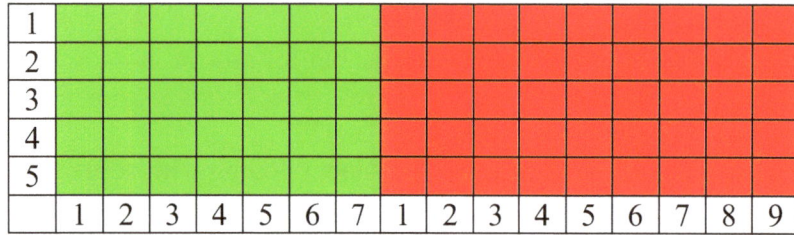

We have two ways to calculate the area of the tilled region, and the two ways produce the same number. How many square meters does the farmer till for me? Sarah.
Sarah: He tills $5 \times 7$ or 35 square meters.
Mentor: And for the neighbor?
Sarah: He tills $5 \times 9$ or 45.
Mentor: How many square meters total does the farmer till?
Sarah: We simply add the areas: $5 \times 7 + 5 \times 9$.
Mentor: The farmer does not know where the line between the green and red regions lies. He wants to get paid for the total job. What calculation does he make?
Sarah: He sees the plot as 5 meters by $7 + 9$ meters. So he multiplies $5 \times (7 + 9)$ to get the total area.
Mentor: What does our formula claim?
Sarah: That $5 \times (7 + 9) = 5 \times 7 + 5 \times 9$.
Mentor: Looking more closely at this formula, we see that multiplying 5 times the sum of the two lengths, $7 + 9$, gives the same result as multiplying 5 times each of the numbers in the sum, $5 \times 7$ and $5 \times 9$, and adding them. Once again, we are expressing the equality of a product and a sum, just as with the difference of two squares, we had expressed the equality of a product and a difference. We express this important formula in words: <u>The product of a number and the sum of two numbers equals the sum of the products of the number and each summand</u>.

A **summand** is one of several numbers being added. Do you think we can extend this formula to have three summands? How can we express the following product as a sum?
$$5 \times (7 + 9 + 4)$$
Paul: I think the second part of the formula will be
$5 \times 7 + 5 \times 9 + 5 \times 4$.

Mentor: How can you justify this claim?

Paul: I think your neighbor wanted to add four more meters to his garden, keeping the 5-meter width.

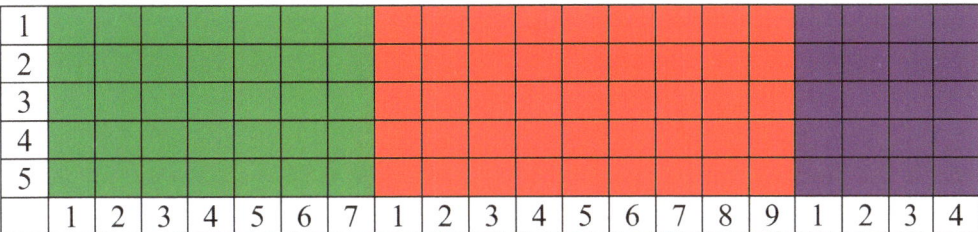

So we can look at this new garden in two ways:
$$5 \times (7 + 9 + 4) \text{ or } 5 \times 7 + 5 \times 9 + 5 \times 4.$$

Mentor: Can you restate our previous formula to cover this situation?

Paul: <u>The product of a number and a sum of three numbers equals the sum of the three products of the number with each summand.</u>

Mentor: What did you have to change in the original formula?

Paul: I changed the 'two' to 'three.'

Mentor: What would you change in your formula to cover the situation where I added more area to my garden, in the same way as my neighbor?

Paul: I would change the 'three' to 'four.'

Mentor: Do you think you could expand the formula again?

Paul: I picture just adding as many additions to your garden as you wish!

Mentor: Excellent. Could anybody see a way to write the following sum as a product:
$$2 \times 7 + 2 \times 3 + 2 \times 5?$$

Sarah: I see this as a garden only 2 meters wide with the other sides being 7, 3, and 5. So I could express the area two ways: $2 \times 7 + 2 \times 3 + 2 \times 5$, and $2 \times (7 + 3 + 5)$.

Mentor: We say that Sarah **factored** the expression $2 \times 7 + 2 \times 3 + 2 \times 5$, by seeing the common factor of 2 in each summand.

We can find less obvious factoring experiences, when given the products as single numbers: $6 + 21 + 33$. Can you spot the common factor, the GCD, and express the sum as a product? George.

George: The GCD is 3 and we can write the sum as $3 \times (2 + 7 + 11)$.

Mentor: Can you identify the GCD of $5 + 7 + 11$? Mary.

Mary: Those are all prime numbers, so 1 is the only common factor!

Mentor: Excellent. How would you write the equation expressing this sum as a product?

Mary: $5 + 7 + 11 = 1 \times (5 + 7 + 11)$.

Mentor: This kind of factoring can appear from time to time, and here is another.

Factor: $5 + 35 + 60$.

Mary: $5 \times (7 + 12)$.

Mentor: Multiply this out to see if it equals the original expression.

Mary: $35 + 60$

Mentor: Where did the 5 at the beginning go?

Mary: Oops. Hmmm. Oh, I see. We have to use 1 again! So now we have
$$5 + 35 + 60 = 5 \times (1 + 7 + 12).$$

Mentor: Pretty sneaky?  In the next Scene, we will expand on our factoring formulas.  But first let me introduce a new piece of mathematical vocabulary, the word term.  In the expression $5 \times 7 + 5 \times 9 + 5 \times 4$ above, we call each of the products ($5 \times 7$, $5 \times 9$, and $5 \times 4$), separated by addition a **term**.  **Terms may be numbers, products, or quotients separated by addition or subtraction.**  In the following expression
$$27 - 3 \times 5 + 12 / 4,$$
we say that 27, $3 \times 5$, and 12/4 are all terms.  The terms themselves do not contain "+" or "–" signs.

Also on the vocabulary theme, note that we call $5 \times 7 + 5 \times 9 + 5 \times 4$ a **sum, because the last operation we perform is an addition** coming after the products have been calculated.  Similarly, we call $5 \times (1 + 7 + 12)$ a **product, because the last operation we perform is multiplication** after we have added the quantity inside the parentheses.

Something new? What?     Some fun? Why?

**Between the Scenes**
1. Factor (or convert the sum of the three given numbers into a product of a common factor and the sum of three smaller numbers): $8 + 10 + 22$.
2. What is the GCD of the three numbers: 30, 54, and 72?  Factor: $30 + 54 + 72$.
3. Factor the sum: $2 + 4 + 6 + 8 + 10$.
4. Factor: $132 + 1067$.
5. Identify the terms in the expression: $4 \times 5 + 32 / 2 - 7$.

**Act 7**     **Scene 5**     **Difference of Two Areas**

Mentor: Now that we can add areas in several ways, let us consider subtracting areas in several forms. Suppose my garden measured 16 meters long by 5 meters wide. How big an area does it cover? Jane.

Jane: The area equals the length, 16 m, times the width, 5 m, or 80 m$^2$.

Mentor: Suppose I now decide the garden demands too much work, so I am going to cut back the length by 7 m, leaving the width unchanged. How do I calculate the area of the remaining garden?

Jane: The area you plan to take out of cultivation measures 5 m × 7 m and equals 35 m$^2$. This you subtract from your 80 m$^2$ to get 45 m$^2$.

Mentor: Your calculations are correct, but we are looking for patterns and need to just indicate the calculations we intend to make. Could you tell us, in terms of 16, 5, and 7 what you just did?

Jane: Yes. 5 × 16 – 5 × 7 gives us the area of the garden after its reduction.

Mentor: As we look at the garden plan below, can you tell us another way to calculate the area.

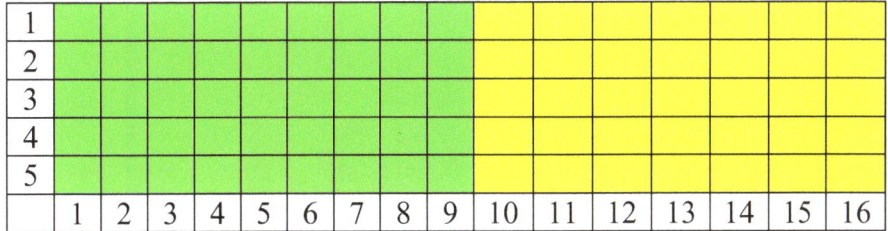

Jane: We could subtract the 7 from the 16 first, and then multiply by 5: 5 × (16 – 7).
Mentor: Does this give us the same area as you told us before?
Jane: Yes. So we can write the equation: 5 × (16 – 7) = 5 × 16 – 5 × 7.
Mentor: What can we say about factoring and subtraction?
Jane: We factor the 5 from the difference the way we factored the 5 from the sum.
Mentor: We were looking for just that pattern! Let's try to factor the following difference: 4 × 17 – 4 × 11. We use the word "term" (defined in the last Scene) to describe any parts of the arithmetical expression separated by a "+" or a "–" sign.
Tom: I see the common factor 4 in each term and can factor the expression to produce
   4 × (17 – 11). So my equation states: 4 × 17 – 4 × 11 = 4 × (17 – 11).
Mentor: Subtraction gives us some new vocabulary. Consider the expression:
   17 – 11 = 6.
We call 6 the **difference** of 17 and 11. 17 we call the **minuend**, the number we diminish by subtracting another number from it. The 11 we call the **subtrahend**, the number we deduct, or subtract, from another. Note, to help our memories, the related words: di<u>mi</u>nish the <u>mi</u>nuend and <u>subtract</u> the <u>subtra</u>hend.

   Name the parts of the following equation: 35 – 21 = 14.
George: 35 is the minuend, 21 the subtrahend, and 14 the difference.
Mentor: 75 – 42 = 33.
Sarah: 75 is the minuend, 42 the subtrahend, and 33 the difference.

Mentor: $3 \times 8 - 3 \times 6 = 3 \times 2$.
Mary: $3 \times 8$ is the minuend, $3 \times 6$ the subtrahend, $3 \times 2$ the difference.

Mentor: Can anybody write the difference of two areas as a product? For instance, the area between two gardens, both 3 yards on a side, might be $3 \times 7 - 3 \times 5$. How could we write this area as a product of the length of the common side and the difference of the two different sides?
Sarah: I see a common multiple of 3 in each product and factor it out, leaving $7 - 5$.
Mentor: Can you write an equation expressing this idea?
Sarah: Sure. $3 \times 7 - 3 \times 5 = 3 \times (7 - 5)$.
Mentor: Can you imagine a case in which you like to do such factoring?
Sarah: If the numbers were big, such as 427 and 425, subtracting one from another might give us smaller numbers to work with, $3 \times (427 - 425)$ or $3 \times 2$ instead of $3 \times 427 - 3 \times 425$.
Mentor: Good. Sometimes we will write the subtraction triple in a column instead of an equation, so $356 - 135 = 221$ becomes:

       356    minuend
       <u>135</u>    subtrahend
       221    difference

Mentor: When we give something a name, it becomes easier to talk about it and ultimately easier to understand. So please think about these names, trying to use them whenever appropriate.

Again, note that the expression $9 - 6 - 1$ requires that we first subtract 6 from 9 to get 3 and then subtract 1 from 3 to get 2. If we subtracted the 1 from the 6 first, to get 5, and then subtracted 5 from 9, we get a different number, namely 4. We could indicate this non-standard order by writing the expression with parentheses: $9 - (6 - 1)$. If we want to indicate the order of the normal subtraction, we could place parentheses around the $9 - 6$, creating the following expression:
$(9 - 6) - 1$. Please note, this only serves to emphasize the normal order of operations. We observe that <u>if three terms are separated by subtraction symbols the order of operations proceeds from left to right</u>.

The addition expression, $9 + 6 + 1$, comes out the same if we first add the $9 + 6$ for 15 and then add 1, to get 16, or add $6 + 1$ to get 7 and add that to 9 for a total of 16. Actually we can speed up our addition if we look for pairs of numbers that add up to 10, such as $9 + 1$, and then add the 6.

**Between the Scenes**
Insert parentheses to indicate the order of operations: $17 \times 4 - 9 \times 3 - 8 \times 2$ .

**Act 7**   **Scene 6**   **Multiplying Two Sums**

Mentor: Four gardeners have decided to share a rectangular piece of land in the following manner. John has the upper left corner, a rectangle 5 m by 9 m; Sue the upper right hand corner, 5 m by 7 m; Henry the lower left corner, 8 m by 9 m; and Willie the lower right hand corner, 8 m by 7 m.

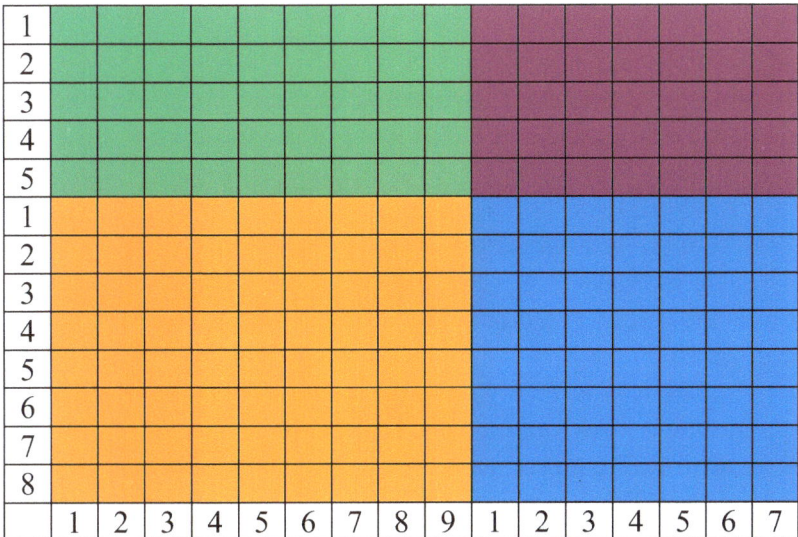

We are looking for two different ways to describe this area. Does somebody have a suggestion? Paul.
Paul: This is a big rectangle 5 + 8 on the vertical side and 9 + 7 on the horizontal side, so we can calculate its area by multiplying the two together.
Mentor: That describes the situation correctly, but how can we express this product without adding the 5 and 8 or the 9 and 7?
Paul: We should use parentheses to make the addition come before the multiplication. Here we have the expression: $(5 + 8) \times (9 + 7)$.
Mentor: Excellent. Can we find another way to calculate the area? Mary.
Mary: Sure. Just add up the areas of the four different colored regions:
   $5 \times 9 + 5 \times 7 + 8 \times 9 + 8 \times 7$.
Mentor: So these two expressions have to be equal. Let's look at the equation and see what patterns we can discover.

$$(5 + 8) \times (9 + 7) = 5 \times 9 + 5 \times 7 + 8 \times 9 + 8 \times 7.$$

For instance, what numbers do we multiply by the 5 in the first quantity?
Sarah: We multiply the 5 times each of the numbers in the other parentheses. We also multiply the 8 times each of the numbers in the other parentheses.
Mentor: What do we multiply by the 9?
Sarah: We multiply each of the numbers in the first quantity by the 9 and the same for the 7!
Mentor: Gage, can you multiply $(6 + 8) \times (2 + 11)$ to give us a sum of products?

Gage: $(6 + 8) \times (2 + 11) = 6 \times 2 + 6 \times 11 + 8 \times 2 + 8 \times 11.$

Mentor: Can somebody tell us how to describe what we are doing? Alexander.
Alexander: <u>To multiply two sums together, multiply each of the summands in one pair of parentheses by each of the summands in the other parentheses and add the four products.</u>
Mentor: We have seen how we can multiply two sums to get a sum of products. Can we reverse this process? If we look at the expression
$$9 \times 5 + 9 \times 3 + 7 \times 5 + 7 \times 3,$$
can we spot a way to turn this sum into a product of two sums?
Tom: I see 9 multiplied times both 5 and 3, so I can factor 9 out of the first two terms, giving me:
$$9 \times (5 + 3) + 7 \times 5 + 7 \times 3.$$
Again, I see a common factor of 7 in the last two terms, so I can factor it:
$$9 \times (5 + 3) + 7 \times (5 + 3).$$
Now I can observe that $(5 + 3)$ is the common factor of the new first and second terms, so I factor out $(5 + 3)$. I get the product:
$$(9 + 7) \times (5 + 3).$$
Mentor: You have found the pattern we were looking for. Good explanation.
We now have a useful tool for transforming a product of sums into a sum of products and the reverse, changing sums of products into products of sums.

**Between the Scenes**
1.  Turn the product $(11 + 4) \times (8 + 13)$ into a sum of four products, drawing the rectangles with the sides indicated in the parentheses.
2.  Factor the sum $6 \times 17 + 6 \times 2 + 19 \times 17 + 19 \times 2$ into a product of two sums, drawing the four rectangles you are adding to get the larger area. Can you do this factoring in more than one way?
3.  A student introduced the author to what her teacher called "Box Multiplication," breaking two two-digit numbers into so many tens and so many units, similar to "Lattice Multiplication." Example: Multiply
$$43 \times 78 = (40 + 3) \times (70 + 8)$$
by adding up the boxes. Perhaps you can do this calculation in your head. $320 + 210 = 530$ and $530 + 2800 = 3330$ and $3330 + 24 = 3354$. Make up your own product and try it.

|  | 40 | 3 |  |
|---|---|---|---|
|  | 2800 | 210 | 70 |
|  | 320 | 24 | 8 |

**Act 7**       **Scene 7**       **Order of Subtractions**

Mentor: We have already seen how <u>a number multiplied by a difference equals the difference of the product of the number with the minuend and with the subtrahend.</u> By way of review, consider the expression $2 \times (10 - 3) = 2 \times 10 - 2 \times 3$. Here we multiply 2 times the difference $10 - 3$ and claim the product equals the difference of $2 \times 10$ less $2 \times 3$. But before we advance, let's look at the vocabulary we have learned. In the expression $2 \times 10 - 2 \times 3 = 2 \times 7$,
what part do we call the minuend, the subtrahend, the difference?
Alexander: $2 \times 10$ is the minuend, $2 \times 3$ the subtrahend, $2 \times 7$ the difference.
Mentor: Pretend I had a garden last summer 19 m long, and 5 m wide. In January I decided to make it 6 m shorter, and in April another 8 m shorter. How would you express this change, starting with the 19 m?
Abbie: $19 - 6 - 8$ m.
Mentor: Tell me the size of my new garden and how you calculated it.
Abbie: $19 - 6 = 13$ and $13 - 8 = 5$. So the new garden's area equals 5 by 5, or 25 m$^2$.
Mentor: By how much have I reduced my garden's length after both cutbacks?
Abbie; You have subtracted a total of $6 + 8$ or 14 m.
Mentor: So what could I have done to calculate the length in a different manner?
Abbie: Subtract the sum of the two lengths you took out of cultivation.
Mentor: What would that formula look like?
Abbie: $19 - 6 - 8 = 19 - (6 + 8)$.
Mentor: How could we put this observation into words?
Abbie: <u>A difference with two subtrahends equals the minuend less the sum of the two subtrahends.</u> More simply, we might say, "Subtract 6 and 8."
Mentor: Well said! If we know we are going to subtract two numbers from a minuend, we can subtract the sum of the two numbers. Calculate $20 - 8 - 7$ two ways. Jane.
Jane: First, I work from left to right, calculating the following:
$20 - 8 = 12$ and then $12 - 7 = 5$.
Second, I add $8 + 7 = 15$, and then subtract 15 from 20 to get 5.
Mentor: So, if my original garden measured 5 m by 20 m, and I decide to decrease the longer side by 8 m and then by 7 m, how do you calculate the resulting area?
Jane: $5 \times 20 - 5 \times 8 - 5 \times 7 = 5 \times 20 - 5 \times (8 + 7)$. Calculating on the left side, I get:
$5 \times 20 - 5 \times 8 - 5 \times 7 = 5 \times (20 - 8 - 7)$, or
$5 \times (12 - 7)$, or
$5 \times 5$.

Calculating on the right side, I get:

$5 \times 20 - 5 \times (8 + 7) = 5 \times 20 - 5 \times 15$, or
$5 \times (20 - 15)$, or
$5 \times 5$.

Mentor: You have shown the two paths nicely.
See something new? Have greater control? Fun?

**Act 7      Scene 8           Multiplying Differences**

Mentor: Just as we drew rectangles to 'see' the formulas for the product of sums, we will continue to pursue the product of differences. We are going to look for two, or more, ways to express the product of two differences. This time we will subtract the areas of smaller rectangles from the area of a bigger rectangle, adding back any rectangles we have taken out twice, thereby making adjustments where necessary. Let us consider the product $(9 - 5) \times (6 - 2)$ as a 9-by-6 rectangle with rectangular parts removed.

The big area we start with measures 9 by 6. We remove the blue-green rectangle, 9 by 2, leaving us the remaining purple-yellow area:
$$9 \times 6 - 9 \times 2.$$
If we remove the yellow-green rectangle, 5 across by 6 down, we have cut the larger rectangle down to the correct size of the purple rectangle,
$$9 \times 6 - 9 \times 2 - 5 \times 6,$$
except for having removed the green rectangle twice! But having removed the green rectangle a second time, we will have to add it back. This area, 5 by 2, we add back in, leaving us with $9 \times 6 - 9 \times 2 - 5 \times 6 + 5 \times 2 = (9 - 5) \times (6 - 2)$.

Remove the parentheses of the expression $(12 - 5) \times (11 - 3)$. Sarah.
Sarah: I draw a picture of the rectangles.

The purple colored rectangle measures $12 - 5$ across the top and $11 - 3$ on the side. We have to remove the other three rectangles from the whole area: $12 \times 11$. I subtract the

107

area of the bottom blue-green rectangle: $12 \times 11 - 12 \times 3$, still leaving me with the yellow rectangle. So I subtract the yellow-green rectangle:
$$12 \times 11 - 12 \times 3 - 11 \times 5.$$
Now I have removed the area of the green rectangle twice, so I have to add back in its area, $3 \times 5$. So
$$12 \times 11 - 12 \times 3 - 5 \times 11 + 5 \times 3 = (12 - 5) \times (11 - 3).$$
Mentor: Good. Reviewing, to calculate the area of a product of differences, subtract the area of the rectangle on the bottom side from the whole rectangle, then subtract the area on the right side, and add back in the rectangle on the bottom right corner. How can we describe the result in terms of the four numbers involved?  The product of two differences equals the product of the two minuends less the product of first minuend and the second subtrahend, less the product of the first subtrahend and the second minuend, plus the product of the two subtrahends. That is a vocabulary problem and needs rehearsal. Drawing the picture seems the easiest approach.

Something new? Some fun solving the puzzle?

**Between the Scenes**
1. Factor the sum: $8 \times 13 - 8 \times 3 - 3 \times 13 + 3 \times 3$. Draw a picture of the rectangles involved in your solution. Label the dimensions!
2. Factor with accompanying figures: $83 \times 71 - 83 \times 59 - 41 \times 71 + 41 \times 59$.

**Act 7    Scene 9    A Difference Times a Sum**

Mentor: We have now had significant experience multiplying sums with sums and differences with differences, so perhaps you can discover how to handle the product of a difference and a sum, such as $(12 - 5) \times (4 + 9)$. How would you start out? Tom.

Tom: I would draw a rectangle, 12 across the top, and $4 + 9 = 13$ down the side.

We want the purple and blue areas, $(12 - 5) \times (4 + 9) = 7 \times 13$, in and the yellow and green out. I add the purple-yellow rectangle, $12 \times 4$, to the blue-green, $12 \times 9$, and subtract the yellow, $5 \times 4$, and green, $5 \times 9$, giving me:

$$(12 - 5) \times (4 + 9) = 12 \times 4 + 12 \times 9 - 5 \times 4 - 5 \times 9.$$

Mentor: By way of summary, we <u>multiply a difference times a sum by multiplying the minuend times each summand and adding the products, and multiplying the subtrahend times each summand and subtracting these products</u>. Now we can multiply two differences or a sum and a difference, but can we reverse this process, taking a sum-difference expression and making it into the product of two differences? Let's work with the expression:

$$13 \times 8 - 13 \times 3 - 6 \times 8 + 6 \times 3.$$

How can we approach this complicated expression?

Alexander: I would start making rectangles. The first two would be $13 \times 8$ and $13 \times 3$, with the $13 \times 3$ rectangle, in blue, being inside the larger one.

|    | 1 | 2 | 3 | 4 | 5 | 6 | 7 | 8 |   |
|----|---|---|---|---|---|---|---|---|---|
| 1  |   |   |   |   |   |   |   |   |   |
| 2  |   |   |   |   |   |   |   |   |   |
| 3  |   |   |   |   |   |   |   |   |   |
| 4  |   |   |   |   |   |   |   |   |   |
| 5  |   |   |   |   |   |   |   |   |   |
| 6  |   |   |   |   |   |   |   |   |   |
| 7  |   |   |   |   |   |   |   |   |   |
| 8  |   |   |   |   |   |   |   |   | 1 |
| 9  |   |   |   |   |   |   |   |   | 2 |
| 10 |   |   |   |   |   |   |   |   | 3 |
| 11 |   |   |   |   |   |   |   |   | 4 |
| 12 |   |   |   |   |   |   |   |   | 5 |
| 13 |   |   |   |   |   |   |   |   | 6 |
|    |   |   |   |   |   | 1 | 2 | 3 |   |

Next, I will take away a 6 × 8 rectangle from the bottom of the largest rectangle, 13 × 8.

|    | 1 | 2 | 3 | 4 | 5 | 6 | 7 | 8 |   |
|----|---|---|---|---|---|---|---|---|---|
| 1  |   |   |   |   |   |   |   |   |   |
| 2  |   |   |   |   |   |   |   |   |   |
| 3  |   |   |   |   |   |   |   |   |   |
| 4  |   |   |   |   |   |   |   |   |   |
| 5  |   |   |   |   |   |   |   |   |   |
| 6  |   |   |   |   |   |   |   |   |   |
| 7  |   |   |   |   |   |   |   |   |   |
| 8  |   |   |   |   |   |   |   |   | 1 |
| 9  |   |   |   |   |   |   |   |   | 2 |
| 10 |   |   |   |   |   |   |   |   | 3 |
| 11 |   |   |   |   |   |   |   |   | 4 |
| 12 |   |   |   |   |   |   |   |   | 5 |
| 13 |   |   |   |   |   |   |   |   | 6 |
|    |   |   |   |   |   | 1 | 2 | 3 |   |

This leaves me with a rectangle (13 − 6) × (8 − 3), if I add the 6 × 3 area to the end of the original expression. Otherwise, I would have removed the 6 × 3 rectangle in the lower right-hand corner twice.

Mentor: Excellent. That was a really tricky factoring problem, because we took away the lower-right rectangle twice! Notice that we can use our rectangles to help go from sum-difference to product as well as the reverse. Our next four-term factoring problem has the last two terms subtracted.

Let's try to factor the expression 9 × 5 + 9 × 3 − 7 × 5 − 7 × 3.
George: I see a big rectangle 9 × (5 + 3).

|   | 1 | 2 | 3 | 4 | 5 | 6 | 7 | 8 |   |
|---|---|---|---|---|---|---|---|---|---|
| 1 |   |   |   |   |   |   |   |   |   |
| 2 |   |   |   |   |   |   |   |   |   |
| 3 |   |   |   |   |   |   |   |   | 1 |
| 4 |   |   |   |   |   |   |   |   | 2 |
| 5 |   |   |   |   |   |   |   |   | 3 |
| 6 |   |   |   |   |   |   |   |   | 4 |
| 7 |   |   |   |   |   |   |   |   | 5 |
| 8 |   |   |   |   |   |   |   |   | 6 |
| 9 |   |   |   |   |   |   |   |   | 7 |
|   |   |   |   |   |   | 1 | 2 | 3 |   |

Now I am going to take away two rectangles, one $7 \times 5$, the other $7 \times 3$, from the bottoms of the green and purple colored rectangles, now shown below in red and yellow, respectively. I subtract $7 \times (5 + 3)$ from the $9 \times (5 + 3)$.

|   | 1 | 2 | 3 | 4 | 5 | 6 | 7 | 8 |   |
|---|---|---|---|---|---|---|---|---|---|
| 1 |   |   |   |   |   |   |   |   |   |
| 2 |   |   |   |   |   |   |   |   |   |
| 3 |   |   |   |   |   |   |   |   | 1 |
| 4 |   |   |   |   |   |   |   |   | 2 |
| 5 |   |   |   |   |   |   |   |   | 3 |
| 6 |   |   |   |   |   |   |   |   | 4 |
| 7 |   |   |   |   |   |   |   |   | 5 |
| 8 |   |   |   |   |   |   |   |   | 6 |
| 9 |   |   |   |   |   |   |   |   | 7 |
|   |   |   |   |   |   | 1 | 2 | 3 |   |

At this point we see that the area we have left, the green plus the purple, equals
$$(9 - 7) \times (5 + 3).$$
Mentor: Nice work. You put it all together. Does anybody see an alternative solution?
Jane: If I reorganized the terms of $9 \times 5 + 9 \times 3 - 7 \times 5 - 7 \times 3$ and wrote

$$9 \times 5 + 9 \times 3 - 7 \times 5 - 7 \times 3 = 9 \times 5 - 7 \times 5 + 9 \times 3 - 7 \times 3,$$
so I could see
$$(9 - 7) \times 5 + (9 - 7) \times 3,$$
showing the common factor of $(9 - 7)$ instead of $(5 + 3)$.


Mentor: To recapitulate: <u>The product of a difference and a sum equals the sum of the minuend times the first summand and of the minuend times the second summand less the subtrahend times the first summand and less the subtrahend times the second summand.</u>

Putting such complicated ideas into words demands concentration, but also helps us learn the vocabulary as well as the concept of four-term factoring.

Did you have fun working out this puzzle? Did you learn something new?

**Between the Scenes**
1. Factor: $23 \times 17 + 23 \times 11 - 6 \times 17 - 6 \times 11$.
2. Factor: $79 \times 41 - 19 \times 41 + 79 \times 13 - 19 \times 13$.

**Act 8        Directed Numbers**

**Scene 1        Number Line**

Mentor: In almost all the past calculations I have asked you to make, I have not asked you to subtract a larger number from a smaller. In this act, we will learn a method of dealing with such calculations by viewing numbers as pointing in one of two directions: right for positive numbers and left for negative numbers. This assignment does not cast a negative reflection on left-handed people but gives us a common basis for talking to each other. On the stock market, negative is down and positive is up, but that will not be the basis for class conversations. Below you see a small section of the **number line**, each **tick** mark indicating the location of a natural number to the right of zero, the **origin** in red, and a **negative number** to the left.

Notice that the Cuisenaire rods do not indicate a front end or a back end, no direction at all. So we shall use the Cuisenaire rods to help us understand the number line initially but afterwards think of **numbers as points on a line** rather than as blocks of wood. The numbers are below and to the right of their tick marks.

Up to this point, we have been able to subtract four from seven, 7 – 4, but not the other way around. To calculate 4 – 7, we will start on the number line at zero and place a purple rod with its left end on the origin, its right on the 4.

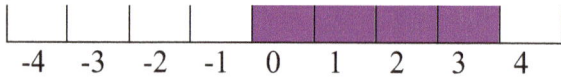

Now, subtracting the seven, we make a parallel train with its right end on the 4-tick mark and extending back to the left.

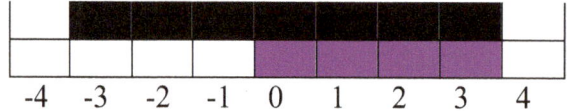

From the number line, we can see that moving from 4 on the right to seven units down on the left we arrive at –3. Note: some calculators have a special symbol, different from subtraction, to indicate the hyphen in –3. In this text we use the same symbol.

So let us now calculate some negative numbers. 3 – 5 = ?
Alexander: I put a green rod on the number line, its left end on the zero tick, its right end extending out to 3. Then I place the yellow rod parallel to the green one, its
right end on the positive 3 and extending off to the left. The left end of the yellow rod reaches -2.

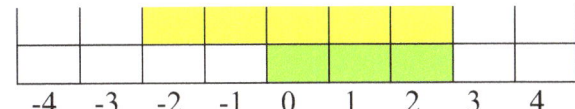

Mentor: Does anybody observe a different way of calculating 3 – 5?
Alexander: Yes. When I want to calculate 3 – 5, I can just calculate 5 – 3, as I always have done, and change the sign.

Mentor: Nice. We do not have to change a lot to use the negative numbers, but there are some changes that may seem strange. If one "-" sign changes the direction of a positive number to a negative number, what will two negative signs do? - -4? Jane.
Jane: You said the sign change indicated a direction change from positive numbers on the right to negative numbers on the left. So why wouldn't a second negative sign simply change the negative number back to the positive one we started with?
Mentor: Of course, that is what it does! - - 10? Tom.
Tom: - - 10 = 10, because we have switched the original positive number from right to left with the first "-" sign and are now changing it back with the second one.

Mentor: Suppose we compare adding a negative number with subtracting a positive one. For instance, consider 3 + -6. What does that give us for a value? Sarah.
Sarah: I think subtracting 6 from 3 would be the same as adding a negative 6 to 3, because, we have the negative numbers reaching from the right end of 3 off to the left.

Mentor: Excellent. We see that <u>adding a negative number yields the same result as subtracting the corresponding positive number</u>: 3 + -6 = 3 – 6. We know we have to learn how to use parentheses in all sorts of situations, and here we consider them with **minus signs** and negative numbers.
    Simplify the expression: -(8 – 3).

Mary: First I draw the number line, placing the brown rod from the origin off to the right.

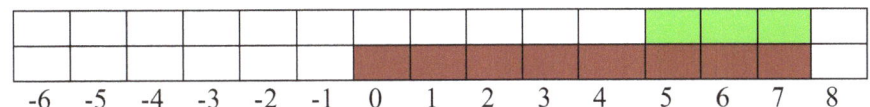

From the right end of the brown rod, I place the green rod, reaching off to the left. From this calculation I know that 8 – 3 = 5. Now I take the yellow rod, reaching from the origin to the left end of the green rod, and turn it around the origin out to the left.

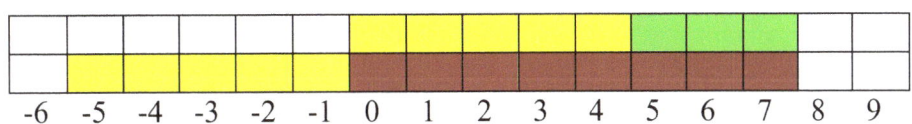

So we can see that –(8 – 3) = -5.

Mentor: Using just 8 and 3 without parentheses, how would you express –5?
Mary: 3 – 8 = -5.
Mentor: We can use this example to illustrate a rule: <u>The negative of a difference equals the difference reversed.</u>  –(8 – 3) = 3 – 8.  Henry, simplify: – (2 – 6).
Henry: I start out with a red rod at the origin, extending off to the right.

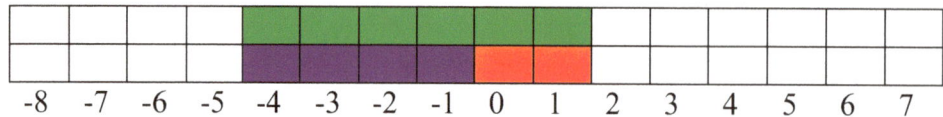

Then from the right end of the red rod I extend the dark green one to the left. This gives us 2 – 6 or negative four. So we have a purple rod from the origin to the left. The minus sign flips the purple rod about the origin to give us plus four.
$$-(2 – 6) = 6 – 2.$$

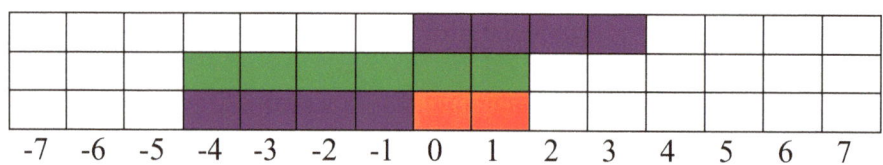

Mentor: Good work. How do we simplify the expression –(-5 – 4)? George.
George: I make a train with yellow and purple rods, extending to the left from the origin.

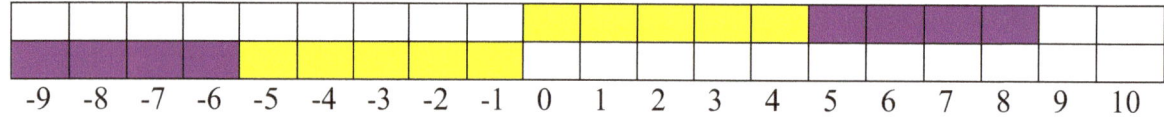

So we have –5 – 4 = -9 which we flip about the origin, to give us 9 or 5 + 4. We can say that <u>the negative of a negative number less a positive number equals the sum of the two corresponding positive numbers.</u>
Mentor: What signs do we see changing when we simplify –(some sum or difference)?
George: Both the signs change.
Mentor: We have a good start on the negative numbers and the number line. Next we have to look at multiplication and division as well as addition and subtraction.
Learn something new? Have some fun?

**Act 8    Scene 2    Combining Negative Numbers**

Mentor: We have seen how we can add a positive and negative number, but we have not considered several other important possibilities. One of these includes adding two negative numbers: -3 + -5. What should we do with these? George.
George: Start with the green rod's right end on the origin and reaching off to the left. Put the yellow rod on the left end of the green one and extend it off to the left. This will give us –8 as a result.

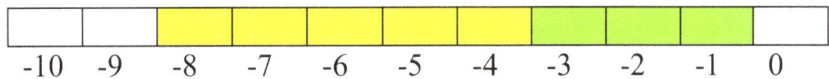

Mentor: Once again, we ask how your calculations will change. What will remain the same?
George: We could have just added 3 and 5 to get 8 and then change the sign to come up with –8. There is relatively little we have to change to add and subtract these strange negative numbers.
Mentor: You have learned how to use parentheses, recognizing
$$(3 + 4) + 5 = 3 + (4 + 5)$$
because adding the first two terms first or the last two first will give the same result. We call this the **associative property**. We will need to use them often when we start multiplying negative numbers. Consider $5 \times (-3)$. What should the result be and why?
Sue: We have learned that multiplication is just addition, so $5 \times (-3)$ equals the sum of five –3's: -3 + -3 + -3 + -3 + -3. The result equals the usual $5 \times 3$ with a minus sign in front: -15.
Mentor: So if we multiply a positive integer times a negative integer, the result is a negative integer, the same integer as if we had multiplied two corresponding positive integers, appending a minus sign. That does not seem like such a big change. Does it?
   Now we consider the biggest change, perhaps, the multiplication of two negative integers: $(-5) \times (-3)$. Does somebody want to tell us about the result: positive or negative? Sue.
Sue: When we put two minus signs in front of a number, like - -3, we found the sign changed back to the original positive number. So maybe it works that way here also.
Mentor: You have spotted a similar situation. Let me show you a rather formal way to illustrate that you can trust your intuition. Consider the following expression:

$$(5)(3) + (5)(-3) + (-5)(-3)$$

We use the parentheses without indicating the multiplication to keep the expression as short as possible. We will simplify the expression, working on the left sum first, and then, starting over again, working on the right sum, using the associative property of +.

$$((5)(3) + (5)(-3)) + (-5)(-3)$$

Can you find the common factor in the left-hand sum? What is it? Sue.
Sue: We can write $5 \times (3 + -3)$ in place of $5 \times 3 + 5 \times -3$.

Mentor: Excellent observation. So now we have the following:

$$(5)(3 + {-3}) + (-5)(-3)$$

Sue: We know that $3 + {-3} = 0$, so we have $5 \times 0$ on the left side of the expression.

$$(5)(0) + (-5)(-3)$$

Mentor: Sue, keep going.
Sue: We agree, also, that anything times zero equals zero. This gives us:

$$0 + (-5)(-3)$$

Adding zero leaves just our product: $(-5)(-3)$.
Mentor: Tom, would you show us how to deal with the other sum in the original expression, pointing out the associative property of + in red parentheses?

$$(5)(3) + ((5)(-3) + (-5)(-3))$$

Tom: Yes. I see a common factor of $-3$ in the last two terms. Factoring I get:

$$(5)(3) + (5 + {-5})(-3)$$

Next I see that $5 + {-5} = 0$. So we multiply 0 times $-3$.

$$(5)(3) + (0)(-3)$$

But, zero times anything equals just zero.

$$(5)(3) + 0$$

And adding zero gives no change: $(5)(3)$
Mentor: What have we just done? Sue.
Sue: Assuming the associative property of +, we have shown that $(5)(3) = (-5)(-3)$. What I said before: two minuses equal a plus!
Mentor: We have made great progress. We will take the opportunity to look at this important result again soon.

**Between the Scenes**
1. Evaluate: $(-4)(-5)(-6) + (-7)(-8)$.
2. Evaluate: $(-6)(3 - 8)(-2)^2$.

**Act 8    Scene 3    Areas with Negative Numbers**

Mentor: Before introducing negative numbers, we calculated the areas of several of different sizes of rectangular regions, using only positive numbers. For instance, if my garden measured 5 m by 8 m and I decided to take one meter off the 5 m side and 2 m off the 8 m side, under the old method, we peeled off two side strips that overlapped and had to add the duplicated section back in. Show us how we did this.

Alexander: OK. The area of the new garden equals
$$(5 - 1)(8 - 2).$$
To write this as a sum, we had the whole $5 \times 8$ colored region less two strips:
$$5 \times 8 - 1 \times 8 - 5 \times 2.$$
Here we took away the bottom red–orange strip, $1 \times 8$, and the right-side blue-orange strip, $5 \times 2$. The net effect was to take away the orange $1 \times 2$ rectangle twice, so we added its area back in:
$$5 \times 8 - 1 \times 8 - 5 \times 2 + 1 \times 2.$$
This way we can see from the garden plot the following:
$$(5 - 1)(8 - 2) = 5 \times 8 - 1 \times 8 - 5 \times 2 + 1 \times 2.$$
Mentor: Now if we know how to multiply negative numbers, we can take a different approach. We can rewrite each difference as the sum of the first positive number and the negative second number:
$$(5 - 1) = (5 + -1) \text{ and } (8 - 2) = (8 + -2).$$
Our rule told us that we multiply each term in the one quantity times each term in the other quantity and add them all together. This gives us the following:
$$(5 + -1)(8 + -2) = 5(8) + 5(-2) + (-1)8 + (-1)(-2)$$

What do we do with all the negative expressions? Jane.
Jane: $5(-2) = -5(2)$ and $(-1)8 = -1(8)$, and $(-1)(-2) = 1(2)$. Putting all this together, we have:
$$(5 - 1)(8 - 2) = 5(8) - 5(2) - 1(8) + (1)(2).$$

Mentor: Why do we get the positive last term now?
Jane: We had to multiply two negatives, producing a positive: $1(2)$. This agrees with our gardening solution.
Mentor: Excellent. We are learning how negative numbers can be used in some simple applications.

**Act 8          Scene 4          More Negative Expressions**

Mentor: We have learned that the product of two negative numbers is positive as is the negative of a negative number. So where might we find some expressions with negative numbers in them? Let's try to simplify the expression: $(-2)^2$. Paul.
Paul: $(-2)^2 = (-2)(-2)$ or $2 \times 2 = 4$.
Mentor: $(-5)^3$. Jane.
Jane: $(-5)^3 = (-5)(-5)(-5)$ or $-125$.
Mentor: A possible source of confusion involves expressions like $-3^4$. We apply the exponent before the negative sign, so $-3^4 = -81$. Note the negative despite the even number of factors! $-6^2$. Alexander.
Alexander: $-6^2 = -(6 \times 6)$ or $-36$.
Mentor: Good use of the parentheses. Simplify: $-5^2(4 - 9)$.
Alexander: $-5^2 = -25$, and $4 - 9 = -5$, so the product equals 125.
Mentor: Evaluate: $-3(-4)^2(-2)$.
Sarah: Negative three times negative two equals positive 6. Negative 4 squared equals sixteen, and six times sixteen equals 96.
Mentor: $(-7)(-11)(-13)$.
Mary: The result must be negative, because we are multiplying three negative numbers. We have 7 times 13 equals 91, while 11 times 91 equals 1001, giving us
$-1001$.

**Between the Scenes**
1.  Evaluate: $(-2)^4 \times (-3)(-4)(-5)$ and $-2^4 \times (-3)(-4)(-5)$.
2.  Evaluate: $(-3)^3 \times (-2)(-5)$ and $-3^3 \times (-2)(-5)$.
3.  Factor: 5005 into primes.

**Act 9    Rational Numbers**

**Scene 1    Common Denominators**

Mentor: In Act 5 Scene 1, we learned how to calculate fractional parts of natural numbers, for instance evaluating $\frac{1}{3} \times 12$ by finding three blocks of one color, purple, making a train the same length as 12 (one orange and one red rod). In all our examples, we made sure that the natural number was a multiple of the denominator of the fraction, i.e. 12 is a multiple of 3. In this Act, we will learn how to evaluate expressions like $\frac{1}{3} \times \frac{1}{5}$ in much the same manner. But before we try to multiply two fractions, we will explore the addition and subtraction of fractions. We start with a simple example. If I add $\frac{1}{6}$ to $\frac{5}{6}$, how many sixths do I have?

George: You have $\frac{6}{6}$, by adding the numerators.

Mentor: Yes. What do I get when adding $\frac{1}{2} + \frac{1}{3}$? Can I simply add the numerators?

George: I can't just add the numerators, 1 + 1, the way I did with the two fractions having a common denominator. Clearly $\frac{1}{2} + \frac{1}{3}$ does not equal $\frac{2}{5}$, as $\frac{1}{2}$ alone is bigger than $\frac{2}{5}$.

Mentor: Good observation! Let us begin by considering the various ways we can express the fraction $\frac{1}{2}$ with Cuisenaire rods. For starters, we construct two parallel trains; one white, the numerator; the other red, the denominator. Why does this represent $\frac{1}{2}$?

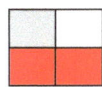

Sarah: We need two white rods to create a train as long as a red rod.
Mentor: So if the numerator train consisted of one red rod, what color denominator rod would you chose?
Sarah: Purple. Purple, because two red rods make a numerator train the same length as the denominator train.

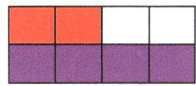

Mentor: Keeping the parallel trains you already have in front of you, tell me what color I would have to select for the denominator train, if I chose a green rod for the numerator.

Jane: You need a dark green denominator train, in order to have the denominator be twice as long as the numerator.

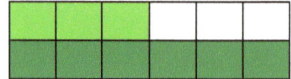

Mentor: Keeping all these train pairs in front of you, let's create train pairs for 1/3. If I pick a white rod for the numerator, what color must I take for the denominator train?
Alexander: Green. Three white rods make a train the same length as a green rod.

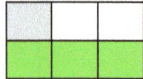

Mentor: Good. Now let us consider a red-numerator train.
Alexander: Dark green; because we have to have three red rods to make a red train the same length as a dark green denominator.

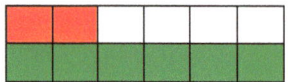

Oh, but we now have a green/dark green representation of $\frac{1}{2}$ and a red/dark green representation of $\frac{1}{3}$. So we can now add $\frac{1}{2}$ and $\frac{1}{3}$ because they both can be written as so many sixths: $\frac{3}{6} + \frac{2}{6}$. All we have to do is add the 3 and the 2. The sum equals $\frac{5}{6}$.

Mentor: Excellent. Looking at what we have just done, what do we see happening to the denominators when we double or triple the numerators?
Alexander: If you double the numerator, then you have to double the denominator to express the same fraction. If you triple the numerator, then you have to triple the denominator.
Mentor: Do you think this process only holds for doubling and tripling?
Alexander: No. Whatever you multiply times the numerator, you will have to multiply times the denominator, or the fraction will change its value.

Mentor: We might state the rule another way. The value of a fraction remains constant, if the same non-zero number is multiplied times both the numerator and denominator. Why have I put in the qualification "non-zero"?
Alexander: Well, if you multiplied zero times both the numerator and denominator, you would have 0/0. I don't know what that would equal.
Mentor: That raises a good question. How do we evaluate 0/0? How do we evaluate 6/2?
Alexander: We know that 6/2 = 3 because 6 = 2 × 3.
Mentor: So, using that logic, we can say 0/0 = ? means that 0 = 0 × ?. The '?' stands for some number, but which one? How do we evaluate zero times any number?
Alexander: 0 times any number equals 0. So any number will do!

Mentor: Exactly. When we divide 6 by 2, we expect to get one specific value, but when we divide by zero, we either find no possible result (6/0 = ? meaning 6 = 0 × ?), or any number (0/0 = 17 meaning 0 = 0 × 17). The first equation, 6/0 = ?, makes no sense, because it means 6 = 0 × ?, but 0 times any number equals 0, and 6 does not equal zero! The second equation, 0/0 = 17, works just fine, because 0 = 0 × 17. The trouble there comes when we observe that not only 17 but also an infinite number of other numbers would do just as well. So we can conclude, that <u>multiplying numerator and denominator by the same non-zero number does not change the value of the fraction.</u>

So how do we find the common denominator of two fractions?

Mary: From this example, all we have to do is multiply the two denominators to create the common denominator.

Mentor: You have made a valid observation. In the next scene, you will find that our practical method requires a more complicated process, but for now your process will do well.

Did you learn something new? Did you have some fun exploring these ideas?

**Between the Scenes**

Using Cuisenaire rods, find the least common denominator and indicated sum of each of the following expressions.

1. $\frac{1}{3} + \frac{2}{5}$.
2. $\frac{1}{2} + \frac{3}{5}$.
3. $\frac{1}{4} + \frac{1}{6}$.
4. $\frac{1}{2} + \frac{1}{4}$.
5. $\frac{1}{2} + \frac{1}{3} + \frac{1}{5}$.

**Act 9     Scene 2     The Least Common Denominator**

Mentor: Most of us find working with small numbers easier than working with larger numbers, so we make an effort to keep our denominators as small as possible. Suppose I wanted to subtract: $\frac{7}{10} - \frac{1}{6}$. What would the possible denominators be?

Jane: I could multiply the top and bottom of the first fraction by 6 and the top and bottom of the second by 10 to get: $\frac{42}{60} - \frac{10}{60}$. That would give me the value $\frac{32}{60}$.

Mentor: What you have done gives a correct answer, but the numbers are larger than necessary. Do 32 and 60 have a common factor we might take out?

Jane: Yes. $32 = 2 \times 16$ and $60 = 2 \times 30$. So we know that $\frac{32}{60}$ equals $\frac{16}{30}$.

Mentor: You have **reduced the fraction** by a common factor of 2. Did you notice another common divisor of $\frac{16}{30}$?

Jane: Yes, I see now that both 16 and 30 are even, so the fraction could be further reduced to $\frac{8}{15}$.

Mentor: Excellent. Going back to the beginning, and recognizing that the Least Common Denominator (**LCD**) of $\frac{7}{10}$ and $\frac{1}{6}$ is 30, not 60, how could we have avoided introducing the first extra 2? What common factor did 10 and 6 possess from the beginning?

Jane: We could just recognize that both are even numbers! $10 = 2 \times 5$ and $6 = 2 \times 3$.

Mentor: If 10 and 6 are going to divide the common denominator, the common multiple will have to be even, but not necessarily a multiple of four. $60 = 4 \times 15$; however, $30 = 2 \times 15$ is all we need. Where did the 15 come from?

Jane: 15 comes from 3 times 5, and 3 is a factor of 6 while 5 is a factor of 10. If the least common denominator is a multiple of 6 it must have a factor of 3; and if it is a multiple of 10, it must be a multiple of 5.

Mentor: You've got it. These numbers, 10 and 6, are small and easy to work with. How could we determine the Greatest Common Divisor (**GCD**) of two numbers we do not immediately recognize? GCD(6,10) = 2. Do you remember how to use the Euclidean Algorithm to find the GCD? Jane.

Jane:     $10 = 1 \times 6 + 4$
          $6 = 1 \times 4 + 2$
          $4 = 2 \times 2 + 0$

So we know that the last non-zero remainder is the GCD of 6 and 10.

Mentor: Calculate the GCD of 75 and 24, using the Euclidean Algorithm method.

Tom:   $75 = 3 \times 24 + 3$
       $24 = 8 \times 3 + 0$

So, we know that 3 is the GCD(75, 24).

Mentor: What good do we get out of this information?

Tom: We can determine all the factors we need to find the LCD of fractions with 75 and 24 in the denominators. 75 = 3 × 25 and 24 = 3 × 8, so the least common multiple of 75 and 24 equals 3 × 25 × 8. We use only one 3 instead of two factors of 3 if we multiplied 75 × 24.

Mentor: Let's add the fractions: $\frac{4}{75} + \frac{5}{24}$.

Tom: I have to multiply the numerator and denominator of $\frac{4}{75}$ by 8 to get the denominator of 600 and likewise the numerator and denominator of $\frac{5}{24}$ by 25. This gives me $\frac{32}{600} + \frac{125}{600} = \frac{157}{600}$.

Mentor: When will the least common multiple equal the product of the two numbers?

Jane: When the greatest common divisor equals one!

Mentor: Note that 4 and 9 are not prime numbers, but their greatest common divisor equals 1, so the LCM(4, 9) = 36. Notice that GCD(4,9) × LCM(4,9) = 4 × 9. What 'rule' would this exemplify? Can you give several more examples?

Also note, when we multiply the numerator and denominator by the same number, say $\frac{8 \times 4}{8 \times 75} = \frac{4}{75}$, we are in effect multiplying by 1, because we know that $\frac{8}{8} = 1$.

**Between the Scenes**

1. Add: $\frac{5}{14} + \frac{2}{21}$.

2. Subtract: $\frac{12}{55} - \frac{5}{33}$.

3. Simplify: $\frac{5}{6} + \frac{4}{9}$.

4. Identify the LCD to calculate the sum: $\frac{32}{143} + \frac{23}{187}$.

5. Calculate: GCD(33, 55) × LCM(33, 55), and 33 × 55.
   Can you guess a general rule? What is it?

**Act 9**       **Scene 3**       **Multiple Additions and Subtractions**

Mentor: We already know how to add $\frac{2}{5}+\frac{1}{3}$, but what changes when we add a third fraction such as $\frac{2}{5}+\frac{1}{3}+\frac{1}{2}$? Alexander.

Alexander: I would add the first two fractions and then add the third fraction to that result. So $\frac{2}{5}+\frac{1}{3}$ has a common denominator of 15, making it the denominator of the sum:

$$\frac{6}{15}+\frac{5}{15}=\frac{11}{15}.$$

Now I will add the $\frac{1}{2}$, so I have a least common denominator of 30.

$$\frac{11}{15}+\frac{1}{2}=\frac{22}{30}+\frac{15}{30}, \text{ or}$$

$$\frac{37}{30}.$$

Mentor: You have done well. Does anybody see the solution another way? Jane.
Jane: I want to find the common denominator for all three fractions at the start. So the least common multiple of 5, 3, and 2 is 30. Then I will change all the fractions into ones with 30 in the denominator: $\frac{12}{30}+\frac{10}{30}+\frac{15}{30}$.

$$\frac{12}{30}+\frac{10}{30}+\frac{15}{30}=\frac{37}{30}$$

Mentor: Fine. Let us consider those two patterns: 1) add the first two fractions and then add the third, or 2) find the common denominator of all three and add the numerators. Try $\frac{9}{14}+\frac{8}{21}-\frac{2}{9}$. Alexander, show us by your method.

Alexander: O.K. $\frac{9}{14}+\frac{8}{21}=\frac{27}{42}+\frac{16}{42}$, or $\frac{43}{42}$. Now I subtract $\frac{43}{42}-\frac{2}{9}$.

$$\frac{43}{42}-\frac{2}{9}=\frac{129}{126}-\frac{28}{126},$$
$$\text{or}$$
$$\frac{101}{126}.$$

Jane: I see the common denominator $126 = 9 \times 14$, so I convert all the fractions to the same denominator.
$$\frac{9 \times 9}{126} + \frac{8 \times 6}{126} - \frac{2 \times 14}{126} = \frac{81}{126} + \frac{48}{126} - \frac{28}{126},$$
or
$$\frac{101}{126}.$$

Mentor: One more challenge. Calculate: $\frac{1}{11} - \frac{3}{5} - \frac{7}{10}$. Sue.

Sue: I see the common denominator as $11 \times 5 \times 2$, and will multiply each numerator by the missing factors:
$$\frac{2 \times 5 \times 1}{2 \times 5 \times 11} - \frac{3 \times 2 \times 11}{2 \times 5 \times 11} - \frac{11 \times 7}{11 \times 10},$$
or
$$\frac{10}{110} - \frac{66}{110} - \frac{77}{110},$$
or
$$-\frac{56}{110} - \frac{77}{110},$$
or
$$-\frac{133}{110}.$$

Mentor: Notice how Sue performed the left-most subtraction first. What would we find if we did the second subtraction first: $\frac{1}{11} - \left(\frac{3}{5} - \frac{7}{10}\right)$? George.

George:
$$\frac{1}{11} - \left(\frac{3}{5} - \frac{7}{10}\right) = \frac{1}{11} - \left(\frac{6}{10} - \frac{7}{10}\right),$$
or
$$\frac{1}{11} - \left(-\frac{1}{10}\right) = \frac{10 + 11}{110},$$
or
$$\frac{21}{110}.$$

Mentor: So, what pattern do we always follow with several subtractions and no parentheses?
George: We calculate from left to right.
Mentor: Do we have another operation that follows this pattern?
George: Yes. Division.

Mentor: Continuing, let's consider how we might calculate and simplify
$$\frac{1}{6} + \frac{3}{4} + \frac{7}{10}.$$

Jane: I would add the first two numbers and then add the third to their sum.
Mentor: Take us through it.
Jane: $\frac{1}{6} + \frac{3}{4}$ we can write as $\frac{2}{12} + \frac{9}{12}$ and add to get $\frac{11}{12}$. Next we add
$$\frac{11}{12} + \frac{7}{10}.$$
The least common multiple of 12 and 10 equals 60, so we multiply 5 times the numerator and denominator of the first fraction and 6 times the numerator and denominator of the second, giving us $\frac{55}{60} + \frac{42}{60}$. Again we note that multiplying numerator and denominator of a fraction by the same number, here by 6, has the same effect of multiplying by 1.

The sum of the numerators equals 97, so we obtain $\frac{97}{60}$. 97 is a prime number, so we cannot reduce the fraction.
Mentor: What do you have to do to show that 97 is prime?
Jane: Check out all the prime numbers up to the square root of 97, or up to 10. We have the list: 2, 3, 5, 7. Because the unit's digit, 7, is not an even number, 2 does not divide 97. Casting away the 9, we see that 3 does not divide 7, so 3 does not divide 97. The last digit is not a 0 or a 5, so 5 does not divide 97. Finally, 7 times 11 equals 77, and 97 − 77 = 20. Since 7 does not divide 20, it will not divide 97.
Mentor: I'm convinced! Does anybody else have a different solution process?
Tom: I would look for the common denominator of all three fractions before adding any of them.
Mentor: OK. Show us how to do it.
Tom: I repeat Jane's fraction work, observing that the least common multiple of 6, 4, and 10 equals 60. Then I rewrite all the fractions: $\frac{10}{60} + \frac{45}{60} + \frac{42}{60}$. This totals up to $\frac{97}{60}$.

Mentor: That was clear. Suppose we were asked to calculate: $\frac{2}{5} - \frac{7}{3} + \frac{1}{6}$.

Sarah: We have to subtract first and then add.
Mentor: Why?
Sarah: Without parentheses, we calculate a mixture of subtractions and additions from left to right; otherwise, we come out with different values. The common denominator equals 30, so I will convert all three fractions to 30ths before subtracting and adding.

$$\frac{2}{5} - \frac{7}{3} + \frac{1}{6} = \frac{12}{30} - \frac{70}{30} + \frac{5}{30},$$
or
$$-\frac{58}{30} + \frac{5}{30},$$
or
$$-\frac{53}{30}.$$

Mentor: What would we have gotten if we added before subtracting?
Sarah: Finding the common denominator will remain the same, but we will add $\frac{70}{30} + \frac{5}{30}$ first.

$$\frac{70}{30} + \frac{5}{30} = \frac{75}{30},$$

that we then subtract from $\frac{12}{30}$.

$$\frac{12}{30} - \frac{75}{30} = -\frac{63}{30},$$
or, reducing we get
$$-\frac{21}{10}.$$

Mentor: Excellent. Can you explain what we have done that we should not have done?
Sarah: Yes. We have subtracted the 1/6, when we should have added it.
Mentor: Let's do one more example. Simplify: $\frac{3}{14} - \frac{5}{21} - \frac{11}{6}$.

Alexander: $\frac{3}{14} - \frac{5}{21} - \frac{11}{6}$ has a common denominator of 42 for all three fractions, so I will rewrite it as follows:

$$\frac{9}{42} - \frac{10}{42} - \frac{77}{42}.$$

Subtracting the $\frac{10}{42}$ from the $\frac{9}{42}$ we get:

$$\frac{9}{42} - \frac{10}{42} - \frac{77}{42} = -\frac{1}{42} - \frac{77}{42}.$$

Now starting at −1 and subtracting 77, we arrive at −78. So we end with:

$$-\frac{1}{42} - \frac{77}{42} = -\frac{78}{42}.$$

But we see that both numerator and denominator are even, so we can reduce the fraction by 2.

$$-\frac{78}{42} = -\frac{39}{21}.$$

Once again we see we can reduce the fraction by 3. This gives us:

$$-\frac{39}{21} = -\frac{13}{7}.$$

Mentor: Earlier, we learned how to add or subtract two fractions, and now we can deal with combined several additions and subtractions in one expression.

Mentor: How can we check with multiplication that these fractions are equal?
Sarah: <u>If the product of the numerator of the first and the denominator of the second equals the product of the denominator of the first and the numerator of the second, then the two fractions are equal.</u> Checking 39 × 7 = 273 and 21 × 13 = 273, we know the fractions are equal.
Mentor: That observation can help us check our work when we simplify fractions. We can say the same thing (two fractions are equal) in a shorter way: <u>The product of the extremes equals the product of the means</u> (the extremes being the numerator of the first and the denominator of the second, the means being the denominator of the first and the numerator of the second). In the next scene we turn to multiplication and division of fractions. Did you find something new and entertaining in this scene?

**Between the Scenes**
1. Using "the product of the means equals the product of the extremes," show that
$$\frac{6}{9} = \frac{14}{21}.$$
2. Evaluate: $\dfrac{7}{10} - \dfrac{8}{15} - \dfrac{1}{6}$.
3. Evaluate: $\dfrac{7}{10} - \left(\dfrac{8}{15} + \dfrac{1}{6}\right)$

**Act 9**  **Scene 4**  **Multiplying and Dividing**

Mentor: In earlier scenes we learned to calculate $\frac{2}{3}$ of 6, or $\frac{2}{3} \times 6$, by finding that rod that is one third as long as 6 and taking two of them. The result equals 4. Suppose now, that we want to find $\frac{2}{3}$ of $\frac{1}{5}$. How could we express $\frac{1}{5}$ as three of something?

Sarah: We have seen that $\frac{1}{5}$ could be written as $\frac{2}{10}$, or $\frac{3}{15}$, or $\frac{4}{20}$, etc. So I will take the $\frac{3}{15}$ form. $\frac{3}{15} = \frac{1}{15} + \frac{1}{15} + \frac{1}{15}$. We see that $\frac{1}{3}$ of $\frac{1}{5}$ equals $\frac{1}{15}$ and we want two of them: $\frac{2}{15}$.

Mentor: If I write this product $\frac{2}{3} \times \frac{1}{5} = \frac{2}{15}$, what do you see for a pattern?

Sarah: We are just <u>multiplying numerator times numerator and denominator times denominator to get the product fraction.</u>

Mentor: We might say: <u>The product of the quotients (fractions) equals the quotient of the products (of the denominators with the denominators and the numerators with the numerators).</u> What is the value of $\frac{4}{3} \times \frac{5}{7}$? Tom.

Tom: I have to divide $\frac{5}{7}$ into three equal parts as

$$\frac{15}{21} = \frac{5}{21} + \frac{5}{21} + \frac{5}{21}.$$

We conclude that $\frac{4}{3} \times \frac{5}{7} = 4 \times \frac{5}{21}$, or $\frac{20}{21}$. This has just the pattern we were talking about: the product of the quotients equals the quotient of the products.

Mentor: How are we to understand division of fractions? Evaluate $\frac{4}{3}$ divided by $\frac{5}{7}$.

Tom: $\dfrac{\left(\frac{4}{3}\right)}{\left(\frac{5}{7}\right)}$ can have 7 multiplied times the numerator and denominator fractions. Then we have

$$\frac{\left(7 \times \frac{4}{3}\right)}{\left(7 \times \frac{5}{7}\right)} = \frac{\left(\frac{28}{3}\right)}{5},$$

because $\frac{1}{7}$ of 7 is one, and five times one equals 5.  At this point, we repeat the process, only that here we are multiplying numerator and denominator by 3 instead of by 7.

$$\frac{3 \times \frac{28}{3}}{3 \times 5} = \frac{28}{15}.$$

Mentor: Tom.  That worked well.  Does somebody see an alternate path?
Mary: Well, we could think of dividing $\frac{28}{3}$ into five equal parts, in other words calculating
$$\frac{1}{5} \text{ of } \frac{28}{3}.$$

So we multiply the numerator and denominator of $\frac{28}{3}$ each by 5, giving us
$$\frac{5 \times 28}{15}.$$

We see that we could write this last expression as we divided the natural numbers with rods.
$$\frac{28}{15} + \frac{28}{15} + \frac{28}{15} + \frac{28}{15} + \frac{28}{15}.$$

So we also obtain
$$\frac{\left(\frac{4}{3}\right)}{\left(\frac{5}{7}\right)} = \frac{28}{15}$$

Mentor: A workable alternative.  If we took all the products apart, into their several factors, what would we find?
Mary: $\frac{4 \times 7}{3 \times 5}$.

Mentor: Can you express this quotient as a product of two fractions?
Mary: Yes. $\frac{4}{3} \times \frac{7}{5}$, following our pattern for multiplication of fractions.

Mentor: Excellent.  What has happened to the $\frac{5}{7}$ in the original denominator?
Mary: It got turned upside down.
Mentor: We might describe the pattern of <u>dividing one fraction by another</u>, saying <u>invert and multiply the denominator (fraction) times the numerator (fraction)</u>.
Let's consider another example to make sure we see how easily we can divide fractions.

Simplify: $\dfrac{\left(\dfrac{7}{8}\right)}{\left(\dfrac{11}{12}\right)}$.

Jane: $\dfrac{\left(\dfrac{7}{8}\right)}{\left(\dfrac{11}{12}\right)} = \dfrac{7}{8} \times \dfrac{12}{11}$, or $\dfrac{7 \times 12}{8 \times 11}$, or $\dfrac{84}{88}$.

Mentor: You have done everything correctly but have left a common factor in numerator and denominator, making the numbers larger than they have to be. Can you spot the common factor?

Jane: Of course, 4 goes into both 84 and 88, so we could reduce this fraction to 21/22.

Mentor: Good. Could you have spotted this common factor earlier in your calculations?

Jane: Yes. We could see in the expression $\dfrac{7 \times 12}{8 \times 11}$ that 4 divides both 8 and 12, reducing them to 2 and 3, respectively. This would give us the fraction $\dfrac{7 \times 3}{2 \times 11} = \dfrac{21}{22}$. This way we would not multiply by the factor of 4 twice only to turn around and divide it out. Also the numbers stay smaller, so they are easier to work with.

Mentor: Did you learn something new? Have some fun discovering these patterns?

**Between the Scenes**

1. A student "simplifies" a fraction $\dfrac{5+7}{2 \times 5} = \dfrac{7}{2}$. Is the result correct? What is the student confusing? (Hint: What is the difference between a term and a factor?)

2. Multiply and simplify: $\dfrac{14}{9} \times \dfrac{12}{49}$.

3. Divide and simplify: $\dfrac{\left(\dfrac{27}{20}\right)}{\left(\dfrac{18}{35}\right)}$.

**Act 9**  **Scene 5**  **Four Operations Together**

Mentor: We have learned how to combine additions and subtractions in longer and more complicated expressions. Now we will throw in multiplication and division as well. Again we recall that multiplication and division come before addition and subtraction unless parentheses change the standard order of operations. Consider the expression:

$$\frac{\frac{3}{5}-\frac{1}{2}}{\frac{7}{10}+\frac{1}{4}}$$

What do we have to do?

Mary: First we have to subtract $\frac{3}{5}-\frac{1}{2}$ and add $\frac{7}{10}+\frac{1}{4}$ and then divide.

Mentor: Take us through those steps.

Mary: $\frac{3}{5}-\frac{1}{2}$ equals $\frac{6}{10}-\frac{5}{10}$ or $\frac{1}{10}$, and $\frac{7}{10}+\frac{1}{4}=\frac{14}{20}+\frac{5}{20}$ or $\frac{19}{20}$. This gives us the fraction:

$$\frac{\left(\frac{1}{10}\right)}{\left(\frac{19}{20}\right)}$$

The next thing we do is to "invert and multiply."

$$\frac{1}{10}\times\frac{20}{19}$$

We know that 10 goes into 20 twice, so we have $\frac{1}{1}\times\frac{2}{19}=\frac{2}{19}$.

Mentor: Excellent. Does any one see an alternative approach?

Sue: If I figure out the least common denominator of all the fractions and multiply numerator and denominator of the larger fraction by it, I will not have to divide one fraction by another.

Mentor: I like the sound of that. Show us what to do.

Sue: We have already found that 20 is the common denominator, so we multiply both top and bottom by 20.

$$\frac{20\left(\frac{3}{5}-\frac{1}{2}\right)}{20\left(\frac{7}{10}+\frac{1}{4}\right)}$$

Simplifying the results, we obtain:

$$\frac{12-10}{14+5}$$

This simplifies down to $\frac{2}{19}$ as we found before.

Mentor: You have shown us an alternative way to simplify a "**complex fraction**," or a fraction with fractions in its numerator or denominator. Perhaps we should try at least one more together to make sure we all understand the options available to us.

Simplify: $\dfrac{\dfrac{3}{2}+\dfrac{11}{10}}{\dfrac{13}{14}-\dfrac{2}{35}}$

Jane: I like to clear out all the denominators first by finding the LCD of all four fractions: 2, 10, 14, and 35. 70 will take care of all the fractions.

$$\frac{70\left(\frac{3}{2}+\frac{11}{10}\right)}{70\left(\frac{13}{14}-\frac{2}{35}\right)}$$

Removing the parentheses, this will simplify to:

$$\frac{35\times 3+7\times 11}{5\times 13-2\times 2}$$

Multiplying we have: $\dfrac{105+77}{65-4}$, or $\dfrac{182}{61}$, and 61 being prime we leave it there.

Mentor: That is fine. Now we need to have one student show the alternative way.

Tom: So I have to add $\frac{3}{2}+\frac{11}{10}$ and subtract $\frac{13}{14}-\frac{2}{35}$ before dividing.

$$\frac{3}{2}+\frac{11}{10}=\frac{15}{10}+\frac{11}{10}, \text{ or } \frac{26}{10},$$

and

$$\frac{13}{14}-\frac{2}{35}=\frac{65}{70}-\frac{4}{70}, \text{ or } \frac{61}{70}$$

Now I have to divide $\frac{26}{10}$ by $\frac{61}{70}$. This I do by inverting the $\frac{61}{70}$ and multiplying:

$$\frac{26}{10}\times\frac{70}{61}.$$

We see that the 10 and the 70 reduce to just 7 in the numerator, leaving us with

$$\frac{26\times 7}{61}, \text{ or}$$

$$\frac{182}{61}.$$

Mentor: I think it's fair to say you can now manage fractions and with practice will be comfortable using them wherever necessary.

**Between the Scenes**

1. Simplify: $\dfrac{\frac{3}{2}-\frac{2}{3}}{\frac{3}{2}+2+\frac{2}{3}}$, observing that $2=\frac{2}{1}$.

2. Simplify: $\dfrac{\frac{1}{2}-\frac{1}{3}}{\frac{1}{4}-\frac{1}{6}}$.

3. Simplify: $\dfrac{\frac{5}{7}+\frac{7}{5}}{\frac{7}{5}-\frac{5}{7}}$.

**Act 10        Exponents, Components**

**Scene 1        Exponents**

Mentor: Earlier we examined the powers of 2 both for a mnemonic for the method of testing if a number is divisible by 2, 4, or 8 and for multiplying using only addition and multiplication by 2.  <u>A positive integer exponent counts the number of times the base occurs in the given expression.</u>  For example: $2^3 = 2 \times 2 \times 2$. Clearly this notation offers a much better way to express the product of fifteen 2's as $2^{15}$ than as $2 \times 2 \times 2 \times \ldots \times 2$.  We have learned to use a useful tool when we learned to use positive exponents.  At this point, we will look at exponents a little more closely, seeing if we can detect some more patterns.  For instance, how many 3's would you expect to find in the expanded version of $3^4 \times 3^6$?

Sarah: I expect to find 10, because there will be four factors of 3 in the first factor and six in the second, so the total will be 10.

Mentor: Excellent.  In the expression $3^4$, we call 3 the **base** and 4 the **exponent**.  If we multiply two exponential expressions with the same base, what would we do with the exponents to combine them?

Sarah: We <u>add the exponents of the factors that have the same base</u>.

Mentor: Suppose we divide one exponential expression by another and both have the same base.  What do we do with the exponents to simplify the expression: $\dfrac{3^5}{3^2}$?

Sarah: We would subtract the exponents.  Why?  Because we can reduce the fraction

$$\frac{3 \times 3 \times 3 \times 3 \times 3}{3 \times 3}$$

by two factors of 3, leaving us with $\dfrac{3 \times 3 \times 3}{1}$ or just $3^3$.

Mentor: Nice.  How do we state a rule for quotients with the same bases?

Sarah: <u>We subtract the exponent of the denominator from that of the numerator in fractions having the same bases.</u>

Mentor: Now several special cases.  What do we mean by $5^1$?

Sarah: That means the product of one five, or just 5.

Mentor: Right!  What do we mean by $5^0$?

Sarah: No fives?

Mentor: Well, that may be one way of looking at it, but consider the fraction $5^3/5^3$.  What does our exponent pattern say we should have for a result, and what should we have for a value?

Sarah: The pattern indicates we should subtract 3 from 3 to get 0, or $5^0$.  However, if we divide 125 by 125, we get 1.  This would lead us to conclude that $5^0 = 1$.

Mentor: Just so.  <u>Any number raised to the zero power will equal one</u>.  Extending this reasoning, how should we understand $7^{-2}$?

Alexander: -49 or –7 squared?

Mentor: Do you mean we should end up with a negative number when we multiply two positive 7's together?

Alexander: No. I guess I am confused.

Mentor: Let us consider the past exponent expressions we have worked with. How about evaluating the expression $\frac{7^3}{7^5}$?

Alexander: That expression equals

$$\frac{7 \times 7 \times 7}{7 \times 7 \times 7 \times 7 \times 7} = \frac{1}{7 \times 7}$$
$$\text{or}$$
$$\frac{1}{7^2}$$

Mentor: Why?

Alexander: Because we can reduce the fraction by the three common factors of seven in both numerator and denominator.

Mentor: What does our pattern of adding or subtracting exponents tell us?

Alexander: 3 – 5 = -2. So $7^{-2} = 1/7^2$. Our pattern then would be that <u>a base number to a negative power is the reciprocal of the base to the positive power</u>, or the power with the negative sign removed.

Mentor: Now that you have made more contact with exponents, we will introduce exponents with exponents in the next scene.

**Between the Scenes**

1.  Simplify: $\frac{2^3 \times 2}{2^5}$.

2.  Evaluate: $17^0 \times 5^2 \times 2^3$.

3.  Simplify, removing all negative exponents: $6^{-3} \times 3^4 \times 5^{-2} \times 10^4$.

**Act 10**  **Scene 2**  **Exponents as Components**

Mentor: In the first scene, we saw that <u>multiplying two numbers with the same base equaled the base with the sum of the two exponents</u> ($5^3 \times 5^4 = 5^7$), and <u>dividing two numbers with the same base equaled the base with the difference of the exponents</u> ($5^3/5^4 = 5^{-1}$). The question arises: what happens when the exponents appear on exponents?

$$5^{4^3} = ?$$

Jane: Usually we evaluate from left to right, as in subtraction and division. So I would raise 5 to the fourth and cube all that:
$$(5 \times 5 \times 5 \times 5) \times (5 \times 5 \times 5 \times 5) \times (5 \times 5 \times 5 \times 5) = 5^{12}$$

This would mean that I should multiply the two exponents together: $3 \times 4 = 12$.

Mentor: You have made a reasonable conjecture, but the system does not work that way. When we raise the exponent to another exponent power, we calculate from right to left! What do we get in this case?

Jane: $4^3 = 4 \times 4 \times 4$, or 64, so our expression equals $5^{64}$. Wow, that expression contains a lot of 5's!

Mentor: Yes. We have to multiply a lot of 5's together to find that value. My calculator tells me that
$$5^{12} = 244140625,$$
and that
$$5^{4^3} = 542101086242752217003726400434970855712890625.$$

So we can see that the way we interpret the expression makes a difference, a big one! We might reasonably ask how many zeros in $10^{64}$?

Jane: That number would equal the product of 64 tens, so 64 zeros. And $10^{12}$ would equal 1 and 12 zeros. Would you like me to write it out?

Mentor: Thank you. That will not be necessary! Now you get some idea of what people are talking about when they say something is **growing exponentially**. If the population doubles every day, or every week, or every decade, then we are going to see big numbers like these. If a population of bacteria starts with 100 members and doubles every day, how many members will it have in ten days?

Sarah: $100 \times 2^{10}$.

Mentor: About how big is $2^{10}$?

Sarah: I know that $2^5 = 32$, so $2^{10} = 32 \times 32$, or 1024.

Mentor: A good way to remember that approximate number sounds like this: <u>Two to the tenth equals (approximately) ten to the third.</u>

Oh, yes. Jane need not be embarrassed that she calculated the original expression with the left-to-right rule, as I have an older calculator that did just that before the manufacturers learned how to program it the correct way.

Did you learn something new? Interesting? Fun?

**Act 10      Scene 3      Difference of Two Cubes & Beyond**

Mentor: We have learned to factor the difference of two squares,
$$7^2 - 3^2 = (7-3)(7+3),$$
using the graphic language of 'square' to see the pattern. Now let us look at the difference of two cubes, $7^3 - 3^3$, finding the first extension of our previous pattern. The figure below represents a cube, 7 units on each side. We calculate its volume by multiplying the length times the width times the depth: $7 \times 7 \times 7 = 7^3$. Because the lines going away from the red 'front' plane do not get closer together the way we see railroad tracks coming together in the distance, the picture may seem to change so you are inside the cube. However, I will describe the figure as having a red 'face' coming out of the page with the other sides going back into the page.

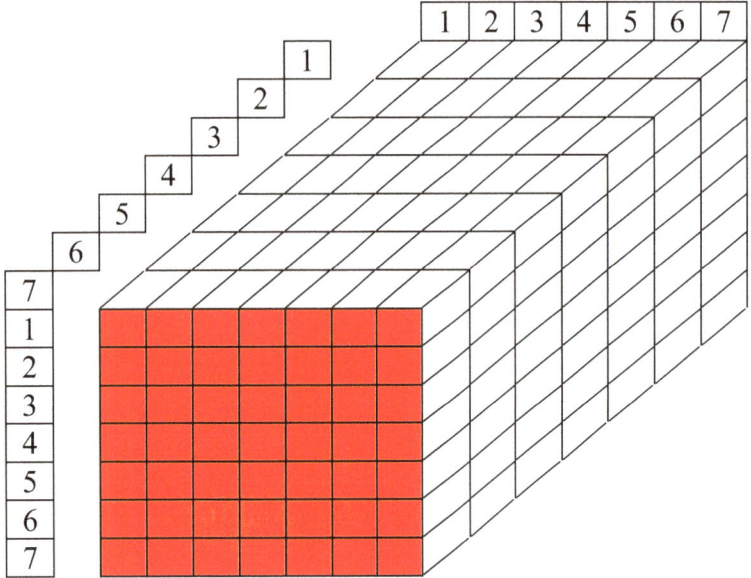

To get our difference of two cubes, we will 'cut' out another cube, $3 \times 3 \times 3$, from the upper right front. After removing this cube, we paint our 'front face' that is left behind red, as shown below.

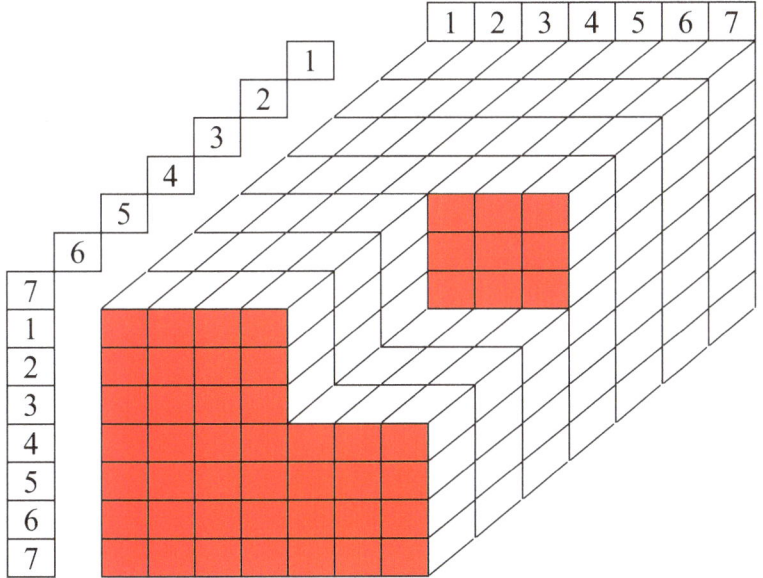

Looking at this figure, we can see the forward-most red surface as the difference of two squares: $7^2 - 3^2 = (7-3)(7+3)$. The new element in our calculations occurs in the figure having a volume, so we must calculate the area of the red surface times the depth of the solid between the red surfaces: $(7-3)(7+3) \times 3$. You can think of adding up the little cubes, $1 \times 1$, on each of the three levels going back into the page:

$$(7-3)(7+3) \times 3 = 40 \times 3, \text{ or}$$
$$120 \text{ little cubes.}$$

Now we slice the front 'L'-shaped solid away from the back part, leaving a solid with dimensions: $4 \times 7 \times 7$, where the $4 = 7 - 3$. Again, we can count the 4 levels of little cubes, finding $7 \times 7$ cubes on each 'face,' for a total of 196 small cubes. Now we add the little cubes in both solids together:

$$\underline{(7-3)} \times (7 \times 7) + \underline{(7-3)}(7+3) \times 3.$$

Plowing our way through all these quantities, we find $(7-3)$ to be a factor of each summand. So we factor the $(7-3)$ out to obtain:

$$\underline{(7-3)}((7 \times 7) + (7+3) \times 3).$$

We multiply by 3 to remove the parentheses around the sum, $(7+3)$, to get:

$$(7-3)((7 \times 7) + (7 \times 3) + (3 \times 3)).$$

Alternatively, using exponents

$$(7-3)(7^2 + 7 \times 3 + 3^2)$$

Mentor: If we look at the trinomial in the factored expression $(7-3)(7^2 + 7 \times 3 + 3^2)$ and write each term with its appropriate exponent, we may see a pattern: $(7^2 + 7^1 \times 3^1 + 3^2)$. Anybody have an idea?
Sue: The sum of the exponents equals 2.
Mentor: Absolutely! How does 2 relate to the original expression?
Sue: 2 is one less than the power of the two cubes.
Mentor: Perhaps if we rewrite the trinomial with 3 and 7 in each term, you may find this other pattern clearer.

$$7^2 \times 3^0 + 7^1 \times 3^1 + 7^0 \times 3^2$$

Tom: I see the exponents of 3 going up from 0 to 2 and the exponents of 7 going down from 2 to 0.
Mentor: Now we have two helpful patterns: <u>in the trinomial factor, the sum of the exponents equals one less than the original exponents and the exponents of the original minuend go down while those of the subtrahend go up.</u> Using this description and the previous example, factor: $7^5 - 3^5$.
Tom: $7^5 - 3^5 = (7-3)(7^4 + 7^3 \times 3^1 + 7^2 \times 3^2 + 7^1 \times 3^3 + 3^4)$.
Mentor: Nicely done. Of course, we do not usually write an expression with the exponent of 1, but in the initial stages we may be helped to determine the pattern. How would we normally write this factorization?
Tom: $7^5 - 3^5 = (7-3)(7^4 + 7^3 \times 3 + 7^2 \times 3^2 + 7 \times 3^3 + 3^4)$.

**Between the Scenes**
1. Factor the expression $8^7 - 3^7$.
2. Factor $11^4 - 5^4$. If we see the exponents 4 as $2^2$, then our factoring may not quite follow the rule we have been following. Why not? Consider $11^{2 \times 2} - 5^{2 \times 2}$.
3. Factor $11^6 - 5^6$. Explain the different possible approaches.
4. Factor $8^{11} - 5^{11}$.
5. Arrange from smallest to largest: $2^{3^4}$, $2^{4^3}$, $4^{2^3}$, $4^{3^2}$.

**Act 10**  **Scene 4**  **Sieve of Eratosthenes**

Mentor: Finding prime numbers can become a chore as we look at larger and larger numbers. Fortunately, most of the numbers, including primes, we have to deal with are relatively small. The sieve of Eratosthenes provides us with a method of listing all the small primes, those less than 100. Take a piece of graph paper, and make a horizontal list from 1 to 10, skipping spaces between numbers. Then, making another horizontal list, continue counting from 11 to 20. Repeat this counting until you reach 100 in the tenth row of numbers. See the table below.

| 1 | 2 | 3 | 4 | 5 | 6 | 7 | 8 | 9 | 10 |
|---|---|---|---|---|---|---|---|---|----|
| 11 | 12 | 13 | 14 | 15 | 16 | 17 | 18 | 19 | 20 |
| 21 | 22 | 23 | 24 | 25 | 26 | 27 | 28 | 29 | 30 |
| 31 | 32 | 33 | 34 | 35 | 36 | 37 | 38 | 39 | 40 |
| 41 | 42 | 43 | 44 | 45 | 46 | 47 | 48 | 49 | 50 |
| 51 | 52 | 53 | 54 | 55 | 56 | 57 | 58 | 59 | 60 |
| 61 | 62 | 63 | 64 | 65 | 66 | 67 | 68 | 69 | 70 |
| 71 | 72 | 73 | 74 | 75 | 76 | 77 | 78 | 79 | 80 |
| 81 | 82 | 83 | 84 | 85 | 86 | 87 | 88 | 89 | 90 |
| 91 | 92 | 93 | 94 | 95 | 96 | 97 | 98 | 99 | 100 |

Now we know that 2 is a prime number, and all the even numbers bigger than 2 are multiples of 2. So we keep 2 and cross out all the even numbers bigger than 2: 4, 6, 8, ... , 100 by turning them orange. This action produces a table looking like this one below.

| 1 | 2 | 3 | 4 | 5 | 6 | 7 | 8 | 9 | 10 |
|---|---|---|---|---|---|---|---|---|---|
| 11 | 12 | 13 | 14 | 15 | 16 | 17 | 18 | 19 | 20 |
| 21 | 22 | 23 | 24 | 25 | 26 | 27 | 28 | 29 | 30 |
| 31 | 32 | 33 | 34 | 35 | 36 | 37 | 38 | 39 | 40 |
| 41 | 42 | 43 | 44 | 45 | 46 | 47 | 48 | 49 | 50 |
| 51 | 52 | 53 | 54 | 55 | 56 | 57 | 58 | 59 | 60 |
| 61 | 62 | 63 | 64 | 65 | 66 | 67 | 68 | 69 | 70 |
| 71 | 72 | 73 | 74 | 75 | 76 | 77 | 78 | 79 | 80 |
| 81 | 82 | 83 | 84 | 85 | 86 | 87 | 88 | 89 | 90 |
| 91 | 92 | 93 | 94 | 95 | 96 | 97 | 98 | 99 | 100 |

Mentor: Do we think of 1 as prime or composite?

Sue: We do not include it in either collection of numbers! We find only one train as long as the white rod, and that is another white rod.

Mentor: Do we still have composite numbers in our list?

Sue: Sure. We have 9, a multiple of 3! And 55, a multiple of 5 and 11.

Mentor: So what should we do next?

Sue: Keep 3 and cross out all its multiples!

Mentor: Excellent. We do it by changing the cell to a red square! Notice we have many numbers that are both multiples of 2 and of 3, but there are many multiples of 3 that we did not cross out when we removed the even numbers. Sue's example of 9 gives us one such number.

| 1 | 2 | 3 | 4 | 5 | 6 | 7 | 8 | 9 | 10 |
|---|---|---|---|---|---|---|---|---|---|
| 11 | 12 | 13 | 14 | 15 | 16 | 17 | 18 | 19 | 20 |
| 21 | 22 | 23 | 24 | 25 | 26 | 27 | 28 | 29 | 30 |
| 31 | 32 | 33 | 34 | 35 | 36 | 37 | 38 | 39 | 40 |
| 41 | 42 | 43 | 44 | 45 | 46 | 47 | 48 | 49 | 50 |
| 51 | 52 | 53 | 54 | 55 | 56 | 57 | 58 | 59 | 60 |
| 61 | 62 | 63 | 64 | 65 | 66 | 67 | 68 | 69 | 70 |
| 71 | 72 | 73 | 74 | 75 | 76 | 77 | 78 | 79 | 80 |
| 81 | 82 | 83 | 84 | 85 | 86 | 87 | 88 | 89 | 90 |
| 91 | 92 | 93 | 94 | 95 | 96 | 97 | 98 | 99 | 100 |

Mentor: What collection of numbers should we cross out this time?

Sue: All the multiples of 5.

Mentor: What about the multiples of 4?

Sue: We crossed those out when we crossed out the multiples of 2 as every multiple of 4 is also a multiple of 2.

Mentor: Now cross out all the multiples of 5 larger than 5. Notice how many numbers we have crossed off. What can we say about most of the numbers that remain?

Sue: They are prime numbers!

Mentor: Note at this point that the first number we have to cross off the list is $5 \times 5$, because we have already eliminated earlier multiples of 5, namely $2 \times 5$, $3 \times 5$, and $4 \times 5$.

| 1 | 2 | 3 | 4 | 5 | 6 | 7 | 8 | 9 | 10 |
|---|---|---|---|---|---|---|---|---|---|
| 11 | 12 | 13 | 14 | 15 | 16 | 17 | 18 | 19 | 20 |
| 21 | 22 | 23 | 24 | 25 | 26 | 27 | 28 | 29 | 30 |
| 31 | 32 | 33 | 34 | 35 | 36 | 37 | 38 | 39 | 40 |
| 41 | 42 | 43 | 44 | 45 | 46 | 47 | 48 | 49 | 50 |
| 51 | 52 | 53 | 54 | 55 | 56 | 57 | 58 | 59 | 60 |
| 61 | 62 | 63 | 64 | 65 | 66 | 67 | 68 | 69 | 70 |
| 71 | 72 | 73 | 74 | 75 | 76 | 77 | 78 | 79 | 80 |
| 81 | 82 | 83 | 84 | 85 | 86 | 87 | 88 | 89 | 90 |
| 91 | 92 | 93 | 94 | 95 | 96 | 97 | 98 | 99 | 100 |

Mentor: What numbers should we cross out next?

Sarah: Multiples of 7 larger than 7.

Mentor: So where should we start, not to be repetitive but also not to miss any composite numbers we should delete?

Sarah: We could start at $7 \times 7 = 49$.

Mentor: Excellent. Let's do it!

| 1 | 2 | 3 | 4 | 5 | 6 | 7 | 8 | 9 | 10 |
|---|---|---|---|---|---|---|---|---|---|
| 11 | 12 | 13 | 14 | 15 | 16 | 17 | 18 | 19 | 20 |
| 21 | 22 | 23 | 24 | 25 | 26 | 27 | 28 | 29 | 30 |
| 31 | 32 | 33 | 34 | 35 | 36 | 37 | 38 | 39 | 40 |
| 41 | 42 | 43 | 44 | 45 | 46 | 47 | 48 | 49 | 50 |
| 51 | 52 | 53 | 54 | 55 | 56 | 57 | 58 | 59 | 60 |
| 61 | 62 | 63 | 64 | 65 | 66 | 67 | 68 | 69 | 70 |
| 71 | 72 | 73 | 74 | 75 | 76 | 77 | 78 | 79 | 80 |
| 81 | 82 | 83 | 84 | 85 | 86 | 87 | 88 | 89 | 90 |
| 91 | 92 | 93 | 94 | 95 | 96 | 97 | 98 | 99 | 100 |

Mentor: Can anybody find a composite number in the list that is left?

Sarah: No. We have learned that we only have to check out numbers up to the square root of a number to be sure there are no factors of it other than one and itself. So we only have to check up to 10 for any number 100 or less. This means that after checking for the primes 2, 3, 5, and 7, we have made all the calculations we must.

Mentor: We call this process the Sieve of Eratosthenes after its inventor because we sift out the unwanted composite numbers and leave only the primes.

     Eratosthenes of Cyrene lived from 276 to 194 BCE. He was the director of the famous library in Alexandria and gifted in all areas of learning. He is most famous for using geometry to calculate the circumference of the earth. (Burton, David M. <u>Elementary Number Theory</u>, pages 52-54)

**Act 10   Scene 5   Divisibility Test for 7**

Mentor: Without using Eratosthenes' sieve, tell me if the number 59 is prime.
B.L.T. Chorus: Yes!
Mentor: Give me evidence.
Berit: We only have to check out 2, 3, 5, and 7.
Mentor: I say 2 divides 59.
Leigh: You are making a mistake, because the last digit must be even for a number to be even.  9 is an odd number.
Mentor: OK, but I claim 4 divides 59.
Tess: If you say 4 divides 59, then 2 has to divide 59 (as we can replace each purple rod with two red ones), but you agreed 2 does not divide 59.
Mentor: So?
Berit: So you contradicted yourself saying 2 divides 59 and 2 does not divide 59.
Leigh and Tess: So 4 does not divide 59.
Mentor: OK, why can we stop checking primes higher than 7?
Berit: We only have to check the prime numbers up to the square root of 59, a number less than 8.
Mentor: I know you know the divisibility properties of 5, so tell me about 7.
Leigh: 7 times 8 equals 56, and seven more will jump up above 59 to 63.  Consequently, we know 7 does not divide 59.

Mentor: Let's look at a divisibility test for 7.  We base the test on the product of $7 \times 11 \times 13$ which equals 1001.  Now we ask if 7 divides 15,359?  We make several observations about 15,359.

$$15{,}359 = 15 \times 1000 + 359$$

Changing the 1000 to 1001, forces us to compensate by subtracting 15.

$$15{,}359 = 15 \times 1001 + 359 - 15$$

Now we know 7 divides 1001, so it divides $15 \times 1001$.  $359 - 15$ gives us all we have to test.  If 7 divides $359 - 15$, or 344, then 7 divides the whole expression and our original number would prove a multiple of 7.  However, in this case we see:

$$344 = 7 \times 49 + 1$$

The remainder of 1 makes this number not a multiple of 7.
Can somebody change one digit of 15,359 to create a multiple of 7?
Tom: Yes.  Try 15,358.

$$15{,}358 = 15 \times 1001 + 358 - 15$$

We know from above that 15 × 1001 has 7 as one of its factors. Now we just have to show 7 divides 358 – 15 or 343. But we have already shown this above, by finding that 344 divided by 7 produces a remainder of 1. So 343 = 7 × 49.

Mentor: Well argued! Now how can we use this devise on a number with many more digits? Consider 7,415,359. We break the number down into blocks of three digits as indicated by the usual commas.

$$7,415,359 = 7 \times 1,000,000 + 415 \times 1,000 + 359,$$

all this without 1001's! Rewriting with the necessary 1001's we get:

$$= 7 \times 1,001,000 - 7 \times 1,000 +$$
$$415 \times 1,001 - 415 +$$
$$359$$

and again compensating for the additional 7,

$$= 7 \times 1,001,000 - 7 \times 1,001 + 7 +$$
$$415 \times 1,001 - 415 +$$
$$359$$

Now we only have to see if <u>7 divides the number, 7 – 415 + 359, breaking the original number into blocks of three digit numbers with alternating signs, starting a positive block on the far right, i.e. under one thousand.</u> 7 divides this number, because 7 divides 7 – 415 + 359, or -49. I will leave it up to you to divide 7 into the given number to check the evidence.

**Between the Scenes**
What other factors can we find using this process? Without dividing the number 7,415,359 by each of these potential factors, determine if they divide the number. Check each result by using division.

**Act 10**   **Scene 6**   **Summing Some Sequences**

Mentor: We have all made stair steps with our Cuisenaire Rods, and now we might look at the sum of such a sequence of numbers: $1 + 2 + 3 + \ldots + 10$.

Alexander: I would just add them up. $1 + 2 = 3$, and $3 + 3 = 6$, and $6 + 4 = 10$, and so on until we add 10 to the rest.

Mentor: Complete the task for us. What value do you get after adding the first ten natural numbers?

Alexander: $10 + 5 = 15$, and $15 + 6 = 21$, and $21 + 7 = 28$, and $28 + 8 = 36$, and $36 + 9 = 45$, and $45 + 10 = 55$.

Mentor: Your calculations give us the right answer, but was this the easiest method?

Mary: I don't think Alexander's method took advantage of the particular numbers he had to add.

Mentor: What do you mean?

Mary: These numbers pair up to make a number of sums equal to 10. I would write the whole list this way: $(1 + 9) + (2 + 8) + (3 + 7) + (4 + 6) + 5 + 10$, giving us four sums of 10, or $40 + 10 + 5$, for 55.

Mentor: Mary has taken advantage of the ease we experience when we can add summands to get 10, and she has shown us how she would do it. Now, suppose we had to calculate the sum of the first 100 natural numbers. Which method would you choose?

Alice: I like Mary's method, because of the easier calculations.

Mentor: Some of us have made rectangles by building upside down stairs on top of right side up stairs. How would you calculate its area?

Alice: The area of the rectangle equals the length times the width, here $10 \times 11$. But we see that we have calculated too much by a factor of 2. We only wanted the area of one set of stairs, so we have to divide by 2: $10 \times 11 / 2$, or $5 \times 11 = 55$. Now I can skip all that adding business, making the sum of the first 100 natural numbers easy: namely $100 \times 101 / 2$, or $50 \times 101 = 5050$.

Mentor: Excellent. You have visualized a pair of very high stairs stacked upside down on top of its twin to make a big rectangle. Which of the three methods would you want to use to calculate the sum of the first thousand natural numbers? Can anybody describe the pattern they see?

149

Sarah: To calculate the sum of the first thousand natural numbers, we multiply one thousand times one thousand plus one and then divide by two. This works because we put the longest rod, 1000, on top of the shortest, 1, to give us a height of 1001 each of 1000 times.

Tom: I see the longest rod, 1000, standing before the first step, and the 999 rod stuck on top of the white rod. This way the height equals 1000, but we have 1001 vertical rods. So the sum equals the same thing.

Mentor: Excellent. Did you learn something new? Did you have fun discovering this method?

**Between the Scenes**

1. How could we calculate the sum of the first 100 even numbers? How do you visualize your calculation? Did learning to factor help you? How many methods do you see available?

2. How could we calculate the sum of the first 100 odd numbers?

3. Calculate the sum of the first 100 multiples of 3. Explain your process. What other numbers could you calculate with your process? Explain!

4. Calculate the sum of the 100 natural numbers from 201 to 300.

# Before and After Questions
# 1-3 Grade Math Teacher Institute

Act 1. Guess the colors of two rods of unequal length in your hands behind your back.

Act 2. Identify and justify your choice of each number as prime or composite: 7, 27, 37.

Act 3. Count: by 7's to 70, by 8's to 80, and by 9's to 90.

Act 4. George divides 20 by 7. What remainder should he get?

Act 5. Evaluate and express in simplest form: $\frac{2}{3} \times 12 + \frac{5}{4} \times 8$.

Act 6. Calculate:  a) the Greatest Common Divisor, and
 b) the Least Common Multiple of 14 and 35.
 c) How are they related?

Act 7. Evaluate in two ways: $(5 + 3^2)(6 + 3)$

Act 8. Evaluate: $(-4)(-7) - (-6)$.

Act 9. Simplify: $\dfrac{\dfrac{5}{6} - \dfrac{2}{21}}{\dfrac{3}{14} + \dfrac{1}{2}}$.

Act 10. Factor:  a) $7^2 - 3^2$
 b) $7^3 - 3^3$.

Bibliography

Few of these books actually gave me material for my text, and some I first saw after *You Can Count On It* was completed. A brief comment should help finding more reference material.

Benjamin, Arthur and Shermer, M., *Secrets of Mental Math: The Mathemagician's Guide to Lightning Calculation and Amazing Math Tricks*, Three Rivers Press, New York, 2006. I read this neat text after I had finished my own, and I am glad I did not see it earlier as I might have tried to use some of Benjamin's ideas. Basically he assumes his students know almost everything in my text. I wanted my young students be able to understand the patterns that Benjamin uses so effectively.

Bowers, Henry and Joan E., *Arithmetical Excursions: An Enrichment of Elementary Mathematics*, Dover Publications, Inc., New York, NY, 1961. This book explains "Lattice Method of Multiplication" and the Doubling Method along with many other arithmetic activities for young children.

Drucker, Peter F., *Managing Oneself*, Harvard Business Review Press, Boston, MA, 2008. Drucker raises the questions about our most effective methods of learning: reader or listener, writer or talker, other activities?

Dudley, Underwood, *Elementary Number Theory*, W. H. Freeman and Company, San Francisco, 1969. This text served as the basis of a course taught at Phillips Academy by my colleague David Penner. I audited the course and mention it for the sake of completeness. It is aimed at an older and more sophisticated population.

Ericsson, K. Anders, *The Road to Excellence: The Acquisition of Expert Performance in the Arts and Sciences, Sports and Games*, Lawrence Erlbaum Associates, Inc. Mahwah, New Jersey 1996. This text describes "Deliberate Practice," for learning a "well-defined task": (1) appropriate difficulty level, (2) informative feedback, (3) opportunities for repetition, and (4) corrections of errors (pages 20, 21).

Kahneman, Daniel, *Thinking, Fast and Slow*, Farrar, Straus and Giroux, New York, 2011. Kahneman shows us how often our lazy intuition is mistaken. Careful System 2 thinking requires effort we frequently don't want to give. It helps to understand his argument to know what the expected value of probability tells us.

Kelley, Tom with J. Littman, *The Art of Innovation*, Doubleday, New York, NY, 2001. Kelley points to the importance of encouraging wide participation without huge negative criticism.

Klein, Gary, *Sources of Power: How People Make Decisions*, MIT Press, Cambridge, MA 1999. Klein gives the example of a math teacher who presents a "core set of examples to serve as analogues" (page 210). His Chapter 4 on The Power of Intuition, points toward the importance of recognizing key patterns (page 31).

Langer, Ellen J., *The Power of Mindful Learning*, Addison-Wesley, Reading, Massachusetts, 1997. I read this book more than a dozen years ago and forgot how much effect it had on me. Langer encourages one to look at alternate methods of solving problems, examining the incorrect solutions, and being open to using ambiguous situations for finding more meaningful insights.

Leonard, George, *Mastery: The Keys to Success and Long-term Fulfillment*, Dutton, New York, 1991. I particularly like his "The Five Master Keys:" Instruction, Practice, Surrender, Intentionality, and The Edge. Part of his first sentence about mastery in the section on instruction states: ...if you intend to take the journey of mastery, the best thing you can do is to arrange for first-rate instruction."

Teaching Company Great Courses
All of these courses are directed toward more experienced students than I had in mind, but the attitudes struck by each Professor suggests the openness and give-and-take I had in mind for my courses.

Benjamin, Arthur, *The Secrets of Mental Math*.

Burger, Edward B., *An Introduction to Number Theory*.

Roberto, Michael A., *The Art of Critical Decision Making*.

Zeitz, Paul, *The Art and Craft of Mathematical Problem Solving*.